Lecture Notebook for Biochemistry

Fourth Edition

Mary K. Campbell
Mount Holyoke University

Shawn O. Farrell
Colorado State University

THOMSON
BROOKS/COLE
Australia • Canada • Mexico • Singapore • Spain • United Kingdom • United States

Printed in the United States of America

2 3 4 5 6 7 05 04 03

ISBN 0-03-034924-9

For more information about our products, contact us at:
Thomson Learning Academic Resource Center
1-800-423-0563

For permission to use material from this text, contact us by:
Phone: 1-800-730-2214
Fax: 1-800-731-2215
Web: www.thomsonrights.com

Asia
Thomson Learning
5 Shenton Way, #01-01
UIC Building
Singapore 068808

Australia
Nelson Thomson Learning
102 Dodds Street
South Street
South Melbourne, Victoria 3205
Australia

Canada
Nelson Thomson Learning
1120 Birchmount Road
Toronto, Ontario M1K 5G4
Canada

Europe/Middle East/South Africa
Thomson Learning
High Holborn House
50/51 Bedford Row
London WC1R 4LR
United Kingdom

Latin America
Thomson Learning
Seneca, 53
Colonia Polanco
11560 Mexico D.F.
Mexico

Spain
Paraninfo Thomson Learning
Calle/Magallanes, 25
28015 Madrid, Spain

To the Student

This **Lecture Notebook for Campbell and Farrell's *Biochemistry,* Fourth Edition** is designed for you to take notes, annotate key figures, and follow along with overhead figures as they are presented in class. The figures in this notebook are the same ones that the publisher has provided to your instructor in overhead transparency form. When your instructor shows one of these figures in your lecture, there is no need to re-draw the art—you can simply make notes on the figures in this book. The pages of this book are perforated and three-hole punched in case you want to tear them out and file them in your own binder.

Table of Contents

Image Number	Figure	Description
93	B.19	Human gene therapy via retroviruses
94	B.23	Steps involved in the construction of a DNA library
95	B.25	Selecting a desired clone from a DNA library
96	B.26	The polymerase chain reaction
97	B.27	The Southern blot
98	B.29	Restriction fragment length polymorphisms
99	B.34	Using a reporter gene
100	12.4	Catabolism and anabolism
101	12.6	NADH and NAD+
102	12.14	Anabolism and catabolism
103	13.2	An aldose and a ketose
104	13.4	Stereochemical relationships among monosaccharides
105	13.8	A comparison of the Fischer, complete Haworth, and abbreviated Haworth structures of α- and β-D-glucose
106	14.1	One molecule of pyruvate is converted to two molecules of pyruvate
107	14.2	The pathway of glycolysis
108	14.3	The first phase of glycolysis
109	14.7	The second phase of glycolysis
110	Table 14.1	The reactions of glycolysis and their standard free energy changes
111	15.5	Glycogen phosphorylase activity is subject to allosteric control and covalent modification
112	15.6	The pathways of gluconeogenesis and glycolysis
113	15.12	The Cori cycle
114	15.15	The oxidative reactions of the pentose phosphate pathway
115	15.16	The nonoxidative reactions of the pentose phosphate pathway
116	15.17	Relationship between the pentose phosphate pathway and glycolysis
117	16.1	The citric acid cycle and catabolism
118	16.3	An overview of the citric acid cycle
119	16.4	The mechanism of the pyruvate dehydrogenase reaction
120	16.7	Control points in the conversion of pyruvate to acetyl-CoA and in the citric acid cycle
121	16.8	The glyoxylate cycle
122	16.9	A summary of catabolism
123	16.11	Transfer of the starting materials of glyconeogenesis from the mitochondrion to the cytosol
124	16.12	Transfer of the starting materials of lipid anabolism from the mitochondrion to the cytosol
125	16.14	A summary of anabolism
126	Table 16.2	The energetics of conversion of pyruvate to carbon dioxide
127	17.7	Respiratory complexes in the inner mitochondrial membrane
128	17.11	A model for the components of ATP synthase
129	17.13	The creation of a proton gradient in chemiosmotic coupling

Image Number	Figure	Description
130	17.21	The glycerol phosphate shuttle
131	17.22	The malate-aspartate shuttle
132	Table 17.1	Standard reduction potentials for biological half-reactions
133	Table 17.2	Energetics of electron transport reactions
134	Table 17.3	Yield of ATP from glucose oxidation
135	18.3	Liberation of fatty acids from triacylglycerols
136	18.5	The role of carnitine in the transfer of acyl groups to the mitochondrial matrix
137	18.6	The β-oxidation of saturated fatty acids
138	18.8	The oxidation of fatty acid containing an odd number of carbon atoms
139	18.10	The oxidation pathway for polyunsaturated fatty acids
140–141	18.15	The first cycle of palmitate synthesis
142–143	18.19	Pathways for the biosynthesis of triacylglycerols
144	Table 18.1	The balance sheet for oxidation of one molecule of stearic acid
145	19.5	Electron flow in Photosystems I and II
146	19.8	Cyclic electron flow coupled to photophosphorylation in Photosystem I
147–148	19.9	Molecular events that take place at the photosynthetic reaction center of *Rhodopseudomonas*
149	19.11	The relationship between photophosphorylation and the proton gradient
150	19.12	The components of the electron transport chain of the thylakoid membrane
151	19.13	The two possible electron transfer pathways in a photosynthetic anaerobe
152	19.14	The main features of the Calvin cycle
153	19.20	The complete Calvin cycle
154	19.21	The C_4 pathway
155	20.5	Families of amino acids based on biosynthetic pathways
156	20.6	The relationship between amino acid metabolism and the citric acid cycle
157	20.17	The urea cycle
158	20.18	The urea cycle and some of its links to the citric acid cycle
159	C.4	Feedback control in the biosynthesis of amino acids of the aspartate family
160	C.9	Histidine biosynthesis
161	C.10	The biosynthesis of prophyrins
162	21.1	Block diagram of intermediary metabolism
163	21.4	Hormonal control system
164	21.5	Activation of adenylyl cyclase by heterodimeric G proteins
165	21.12	A two-stage process leads to the growth and differentiation of T cells

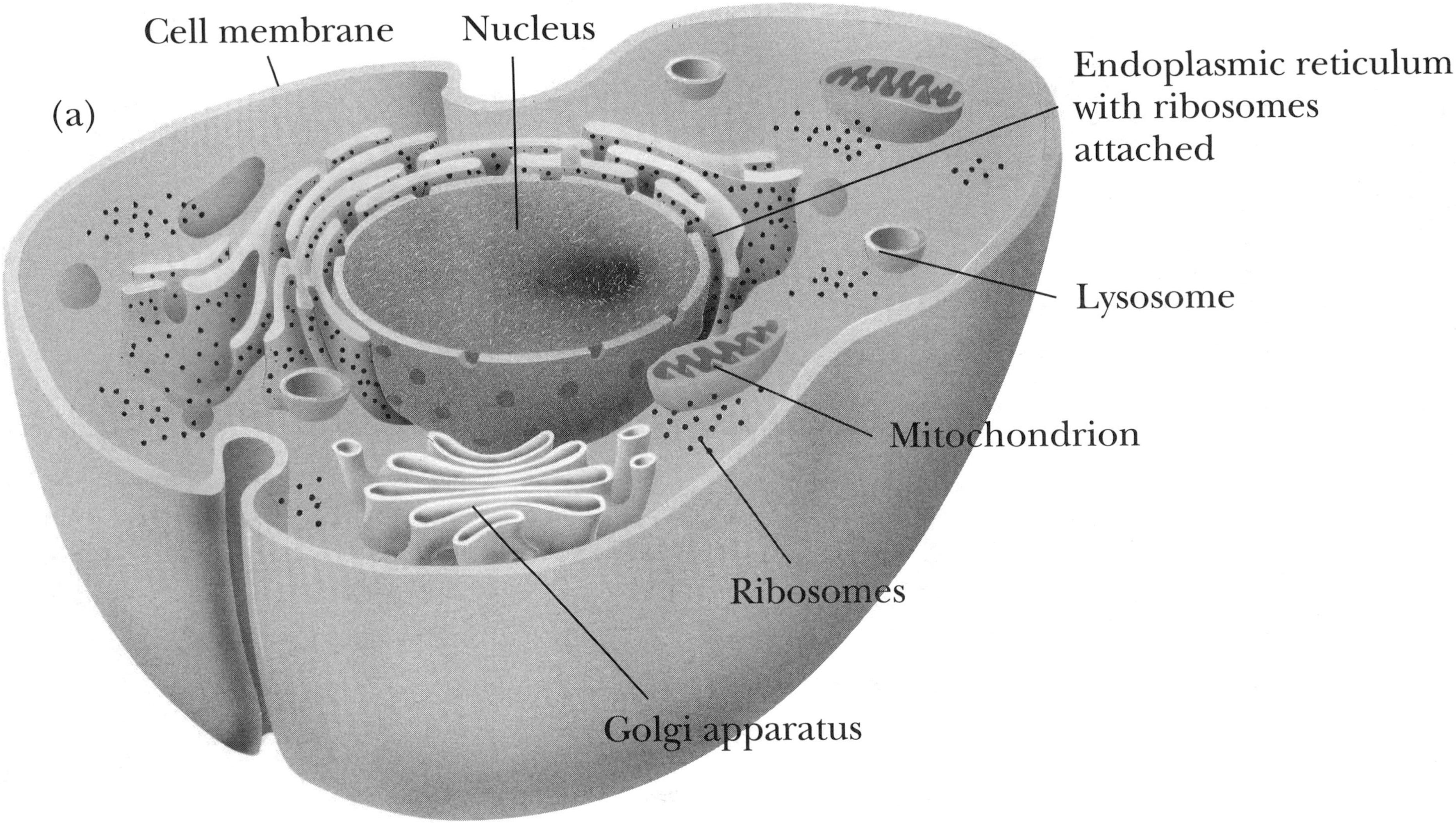
(a)
Cell membrane
Nucleus
Endoplasmic reticulum with ribosomes attached
Lysosome
Mitochondrion
Ribosomes
Golgi apparatus

Figure 1.10a A comparison of an animal cell

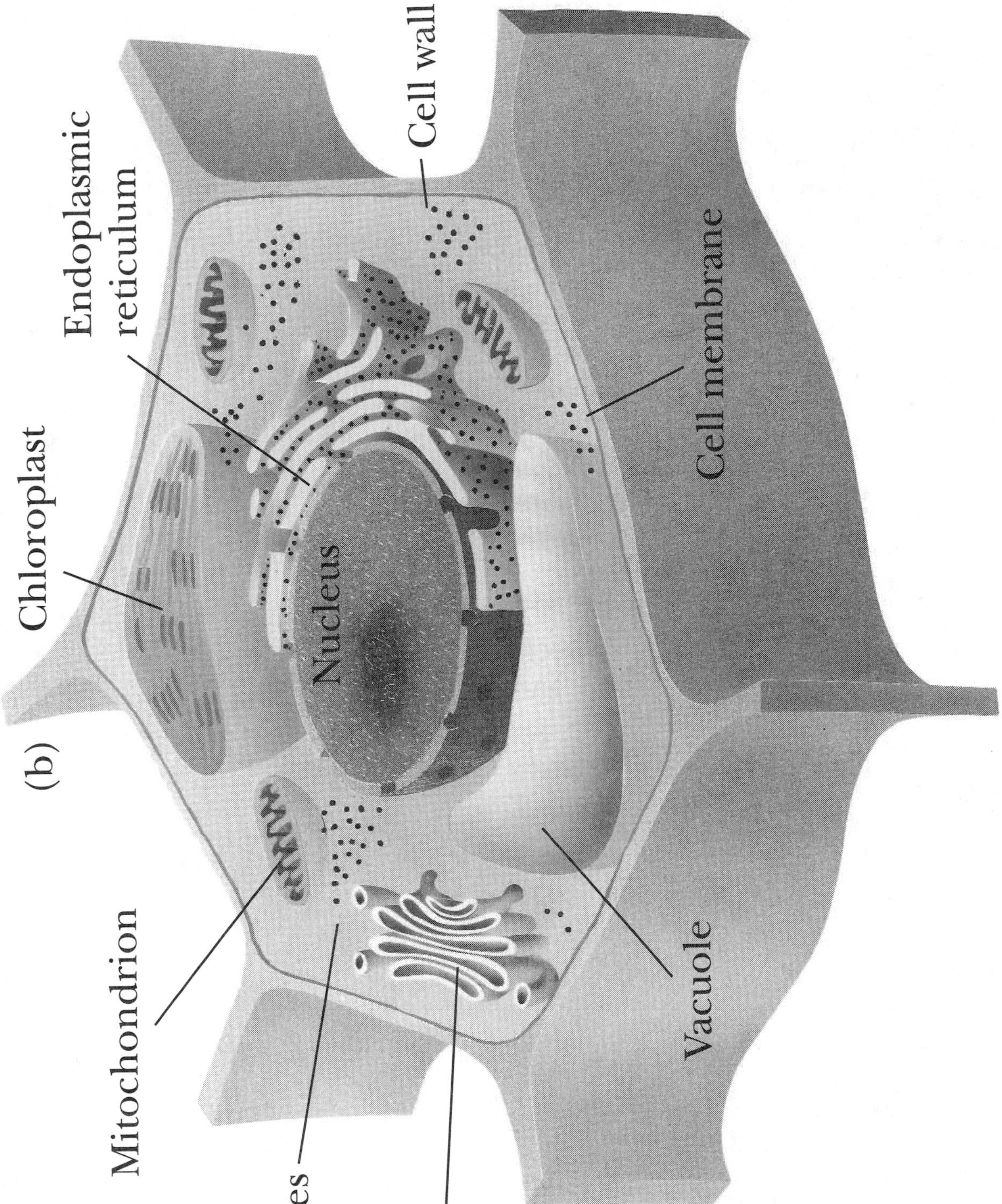
(b)
Chloroplast
Endoplasmic reticulum
Mitochondrion
Ribosomes
Nucleus
Golgi apparatus
Cell wall
Vacuole
Cell membrane

Figure 1.10b A comparison of a plant cell

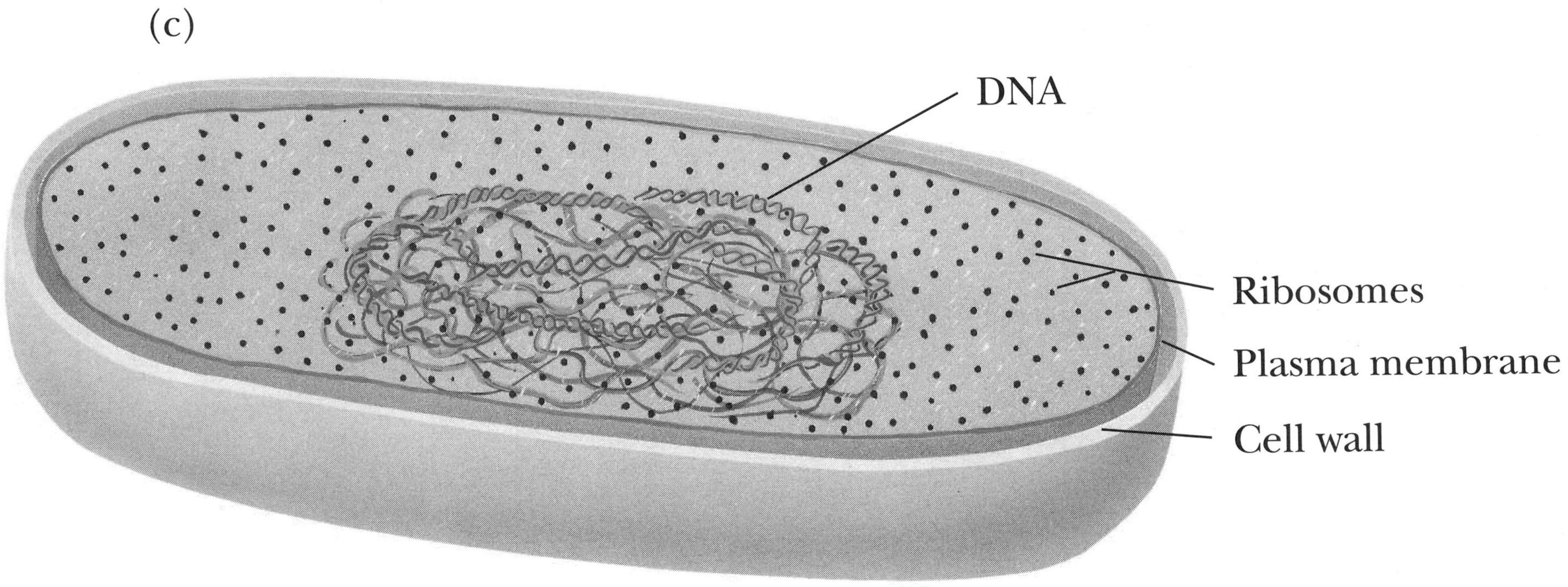

Figure 1.10c A comparison of a prokaryotic cell

Class of Compound	General Structure	Characteristic Functional Group	Name of Functional Group	Example
Alkenes	$RCH=CH_2$ $RCH=CHR$ $R_2C=CHR$ $R_2C=CR_2$	$C=C$	Double bond	$CH_2=CH_2$
Alcohols	ROH	—OH	Hydroxyl group	CH_3CH_2OH
Ethers	ROR	—O—	Ether group	CH_3OCH_3
Amines	RNH_2 R_2NH R_3N	—N<	Amino group	CH_3NH_2
Thiols	RSH	—SH	Sulfhydryl group	CH_3SH
Aldehydes	R—C(=O)—H	—C(=O)—	Carbonyl group	$CH_3C(=O)H$
Ketones	R—C(=O)—R	—C(=O)—	Carbonyl group	$CH_3C(=O)CH_3$
Carboxylic acids	R—C(=O)—OH	—C(=O)—OH	Carboxyl group	$CH_3C(=O)OH$
Esters	R—C(=O)—OR	—C(=O)—OR	Ester group	$CH_3C(=O)OCH_3$
Amides	R—C(=O)—NR_2 R—C(=O)—NHR R—C(=O)—NH_2	—C(=O)—N<	Amide group	$CH_3C(=O)N(CH_3)_2$
Phosphoric acid esters	R—O—P(=O)(OH)—OH	—O—P(=O)(OH)—OH	Phosphoric ester group	CH_3—O—P(=O)(OH)—OH
Phosphoric acid anhydrides	R—O—P(=O)(OH)—O—P(=O)(OH)—OH	—P(=O)(OH)—O—P(=O)(OH)—	Phosphoric anhydride group	HO—P(=O)(OH)—O—P(=O)(OH)—OH

The symbol R refers to any carbon-containing group. When there are several R groups in the same molecule, they may be different

Table 1.1 Functional groups of biochemical importance

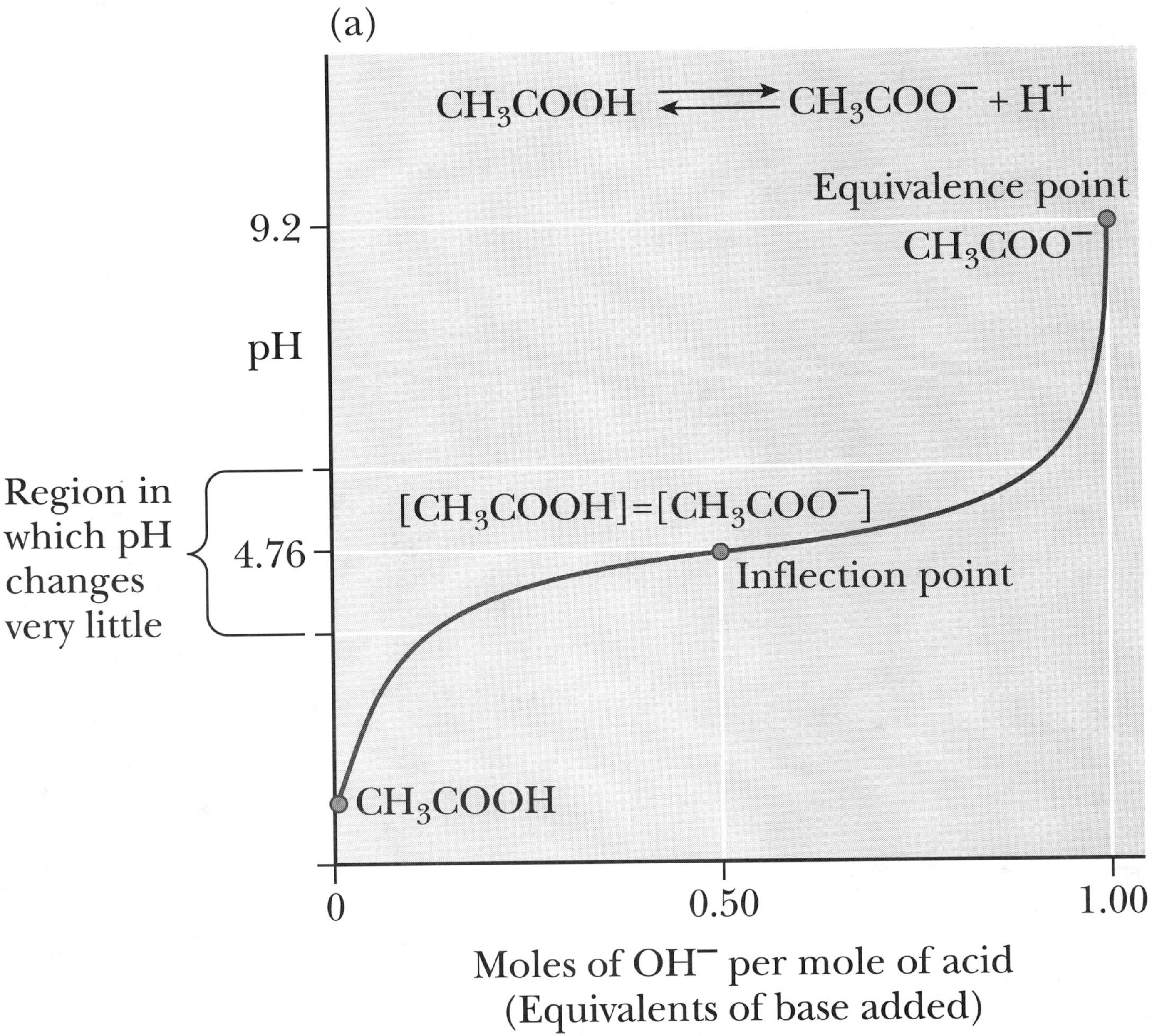

(a)
$CH_3COOH \rightleftharpoons CH_3COO^- + H^+$
Equivalence point
CH_3COO^-
9.2
pH
Region in which pH changes very little
4.76
$[CH_3COOH]=[CH_3COO^-]$
Inflection point
CH_3COOH
0
0.50
1.00
Moles of OH^- per mole of acid
(Equivalents of base added)

Figure 2.9a Titration of acetic acid with NaOH

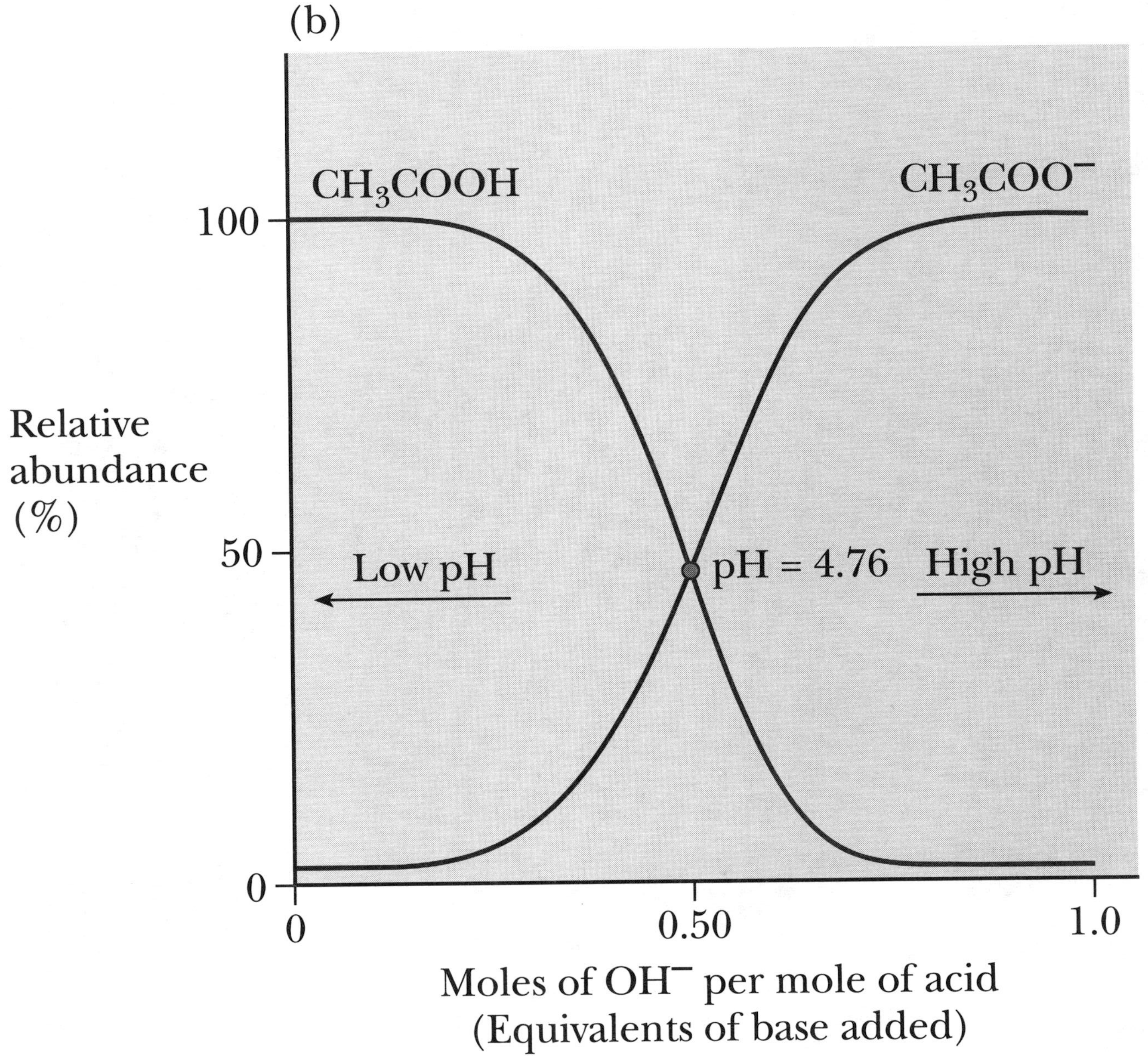

Figure 2.9b Relative abundance of acetic acid and acetate ion during a titration

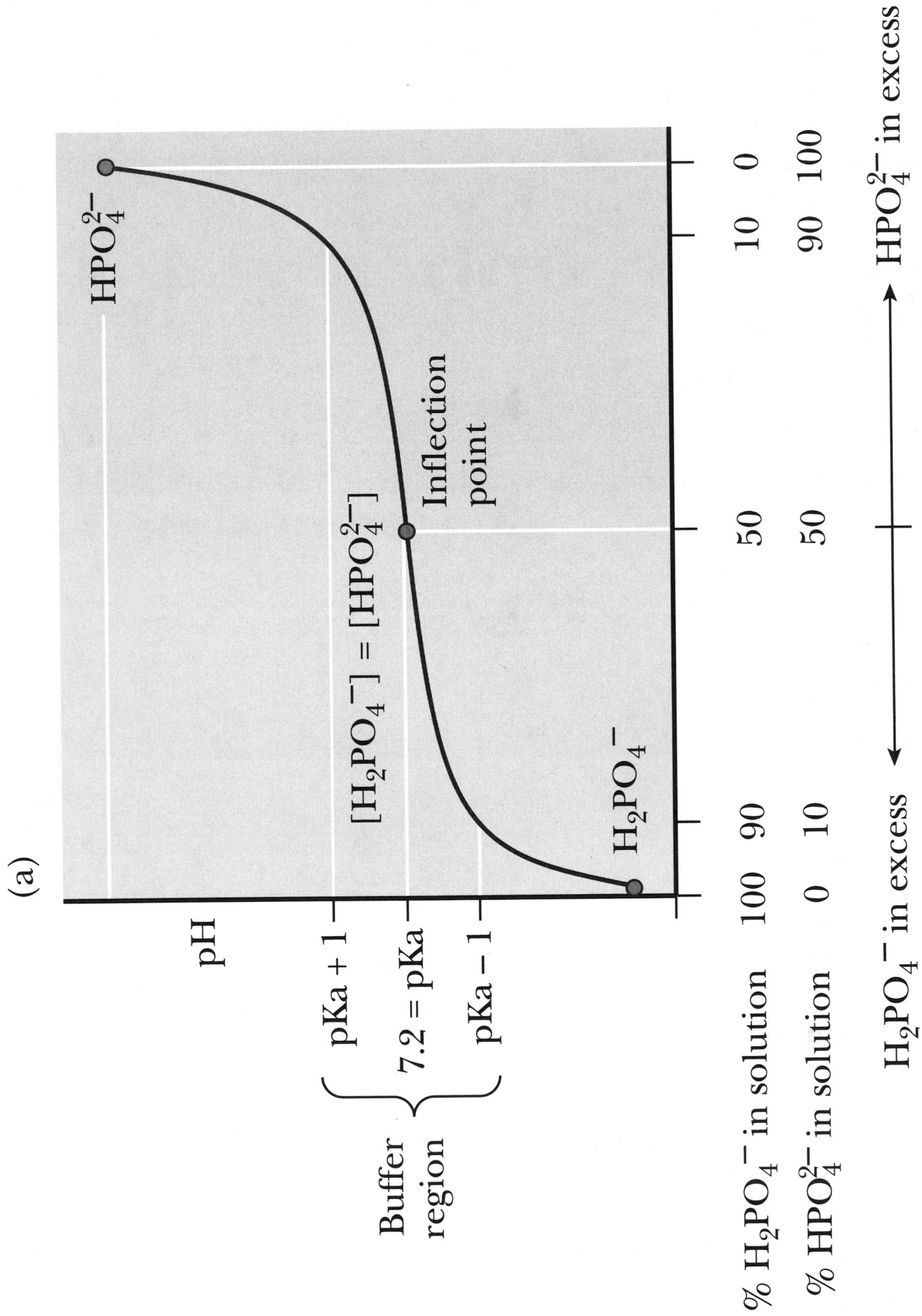

Figure 2.11a Titration curve and buffering action

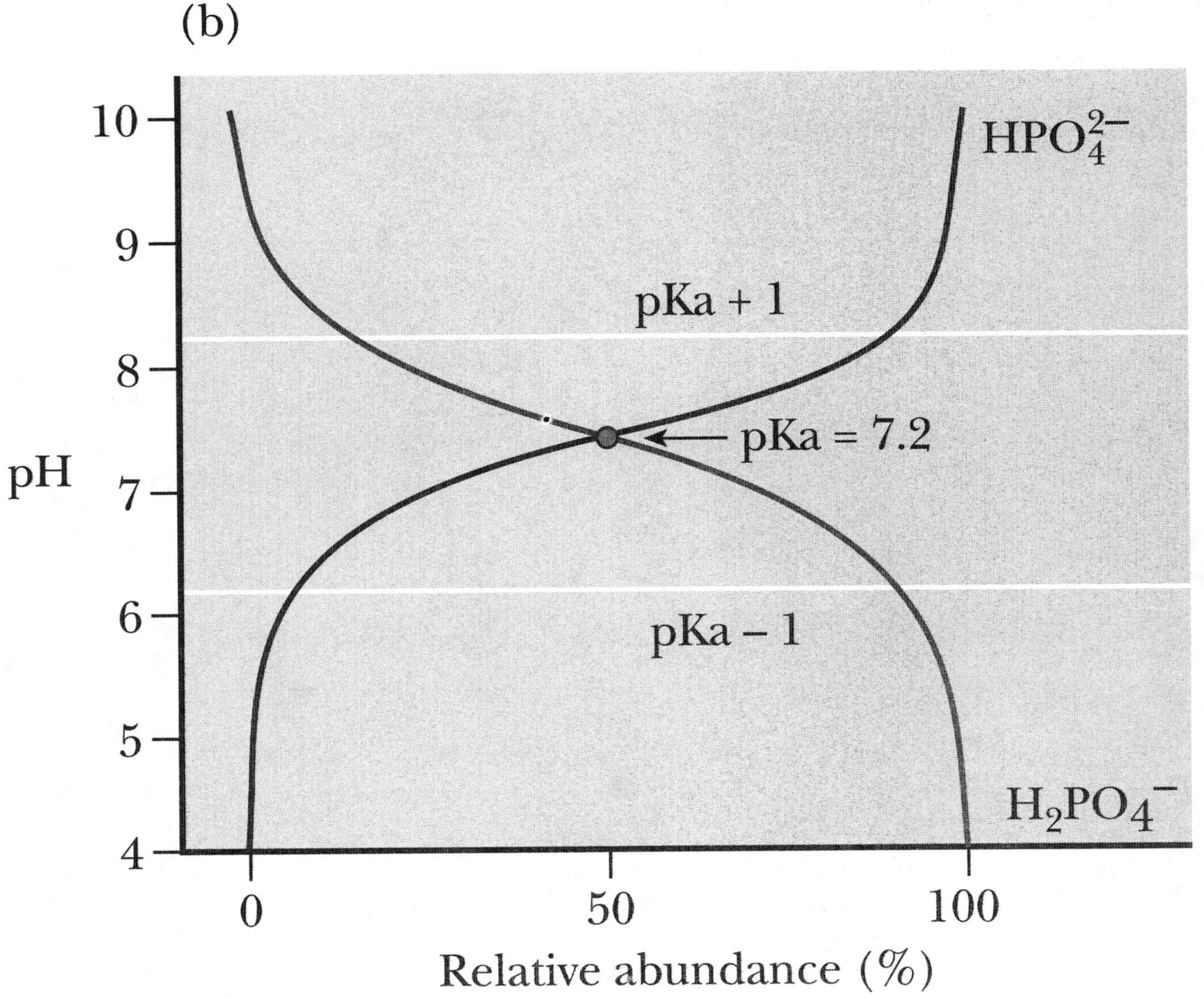

Figure 2.11b Titration curve and buffering action

TABLE 2.8 Acid and Base Forms of Some Useful Biochemical Buffers

Acid Form		Base Form	pK_a
TRIS—H^+ (protonated form) $(HOCH_2)_3CNH_3^+$	N—*tris*[hydroxymethyl]aminomethane (TRIS) ⇌	TRIS (free amine) $(HOCH_2)_3CNH_2$	8.3
$^-$TES—H^+ (zwitterionic form) $(HOCH_2)_3C\overset{+}{N}H_2CH_2CH_2SO_3^-$	N—*tris*[hydroxymethyl]methyl-2-aminoethane sulfonate (TES) ⇌	$^-$TES (anionic form) $(HOCH_2)_3CNHCH_2CH_2SO_3^-$	7.55
$^-$HEPES—H^+ (zwitterionic form) 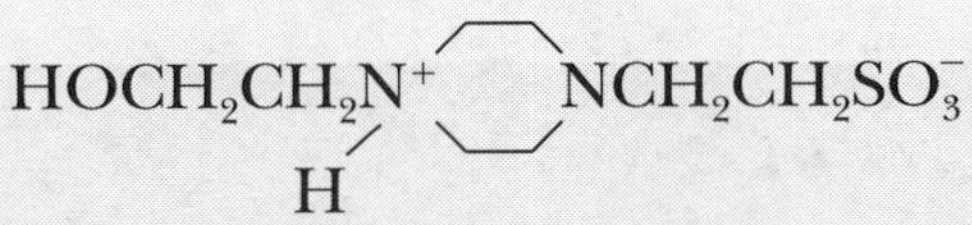	N—2—hydroxyethylpiperazine-N′-2-ethane sulfonate (HEPES) ⇌	$^-$HEPES (anionic form) 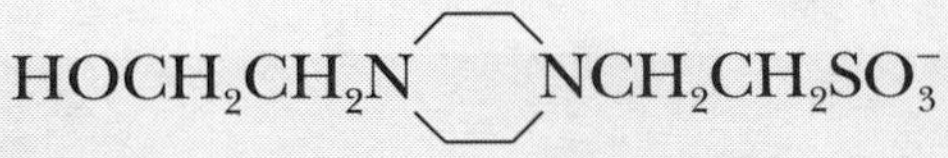	7.55
$^-$MOPS—H^+ (zwitterionic form) 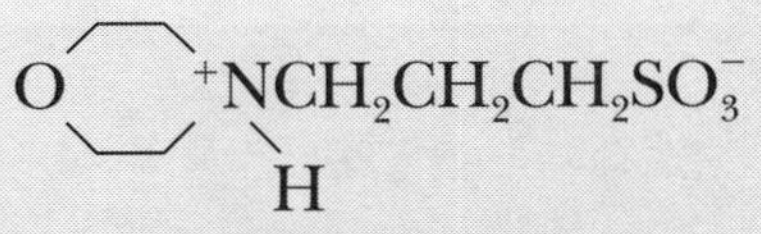	3—[N—morpholino]propane-sulfonic acid (MOPS) ⇌	$^-$MOPS (anionic form) 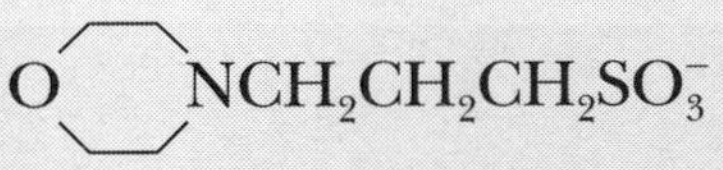	7.2
$^{2-}$PIPES—H^+ (protonated dianion) 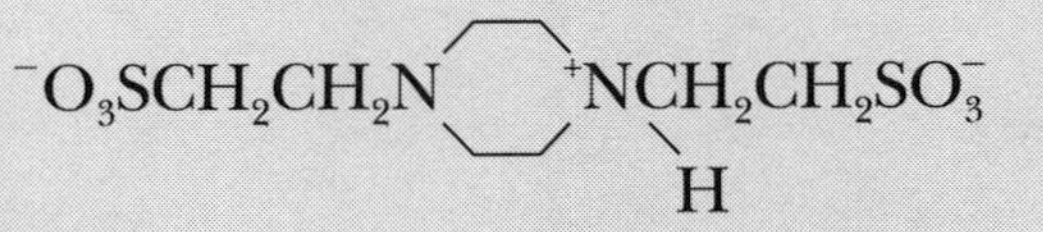	Piperazine—N,N′-*bis*[2-ethanesulfonic acid] (PIPES) ⇌	$^{2-}$PIPES (dianion) 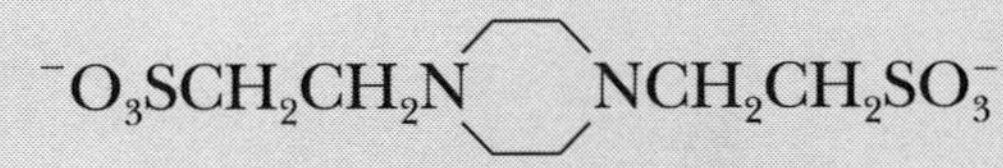	6.8

Table 2.8 Acid and base forms of some buffers

(a)

(b)

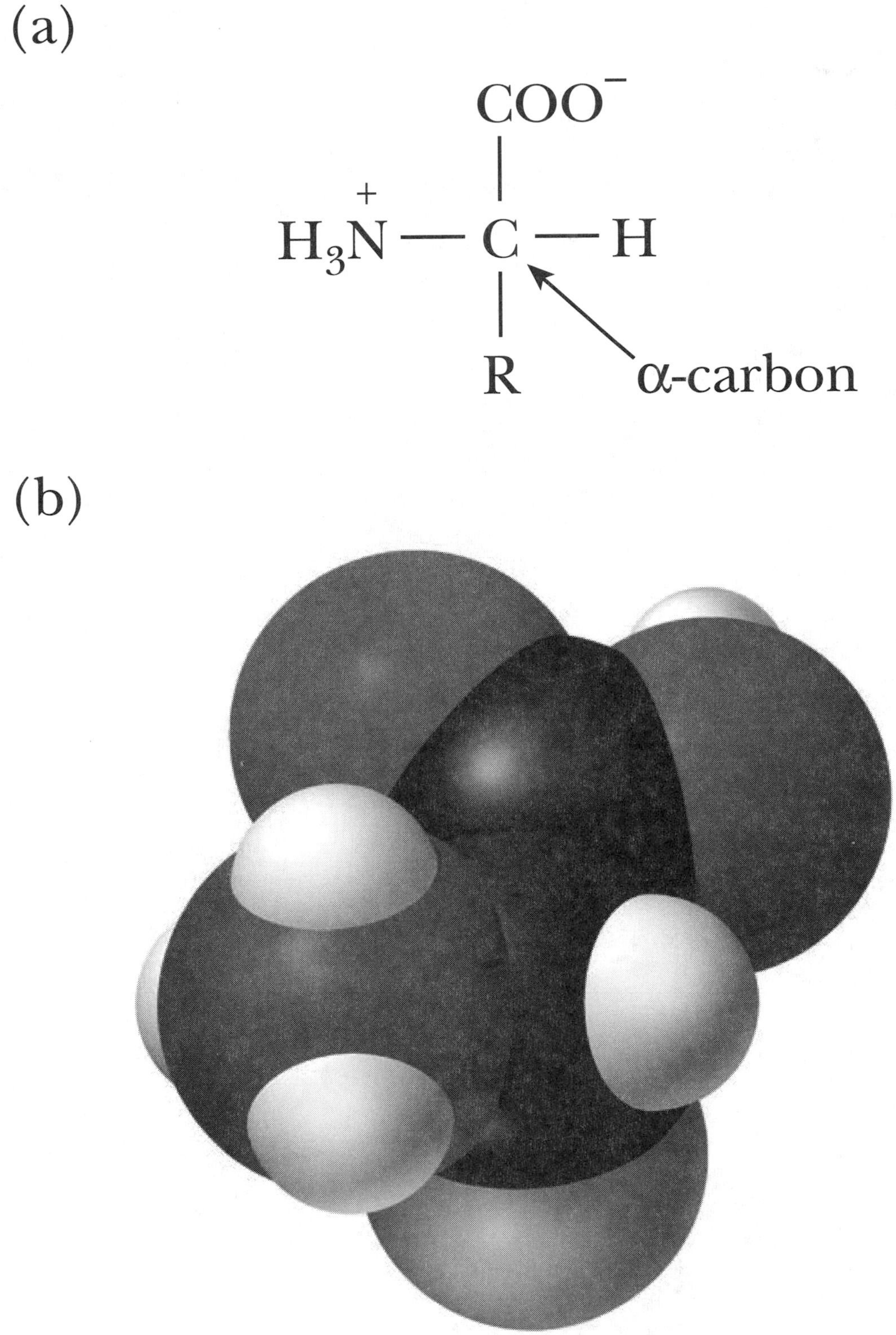

Figure 3.1 Amino acid structure

(a) Non-polar (hydrophobic)

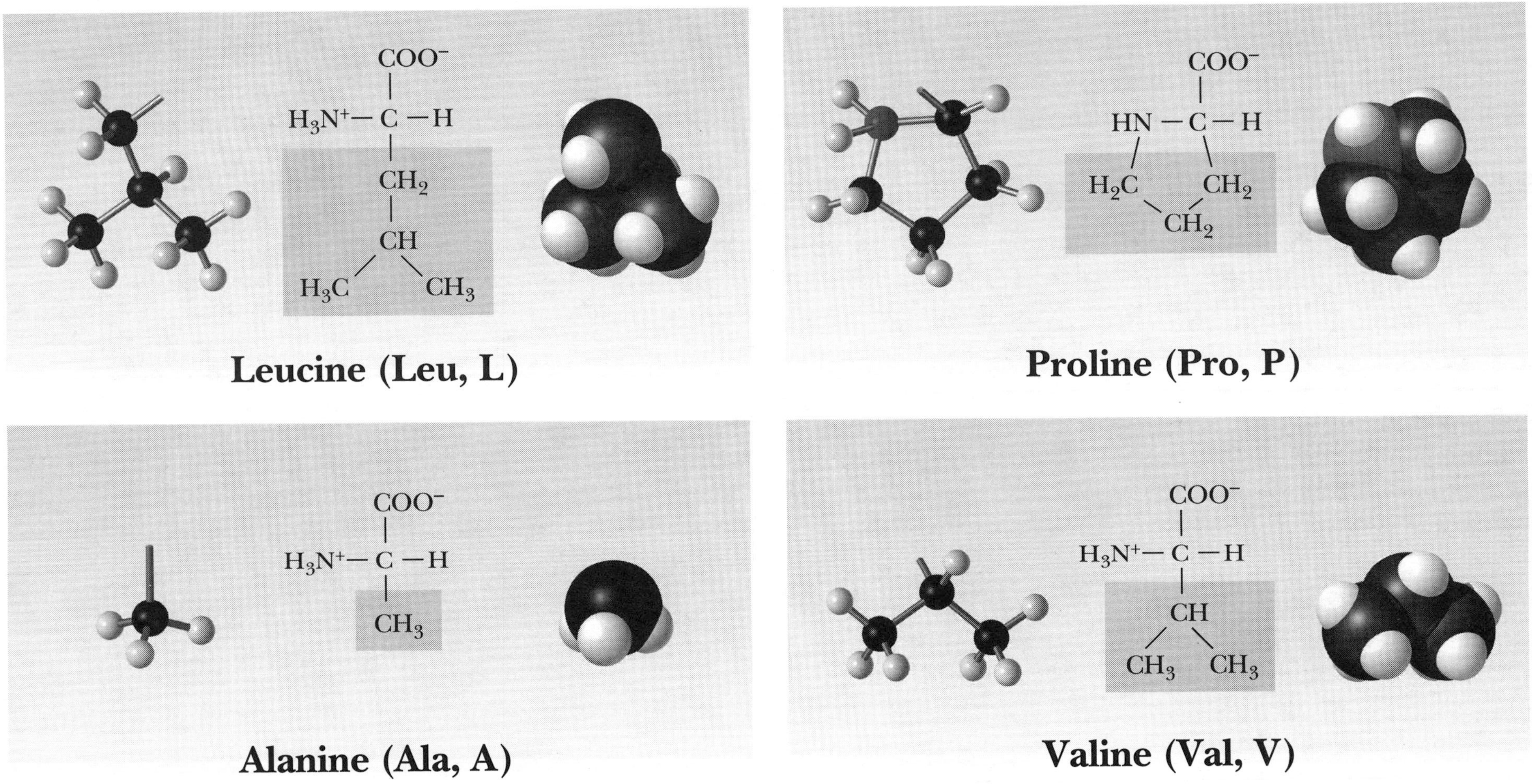

Figure 3.4a Amino acids and their structures

(b) Polar, uncharged

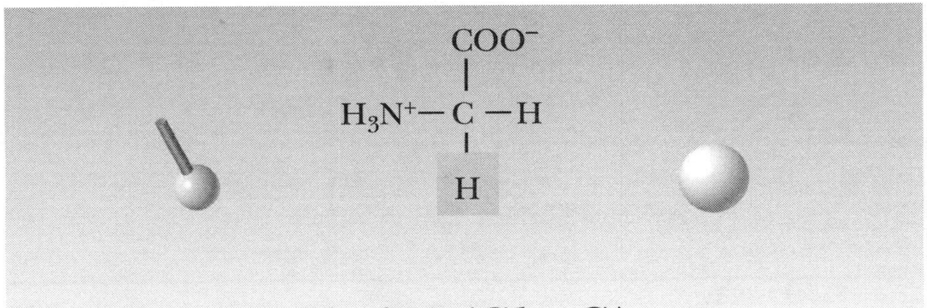

Glycine (Gly, G)

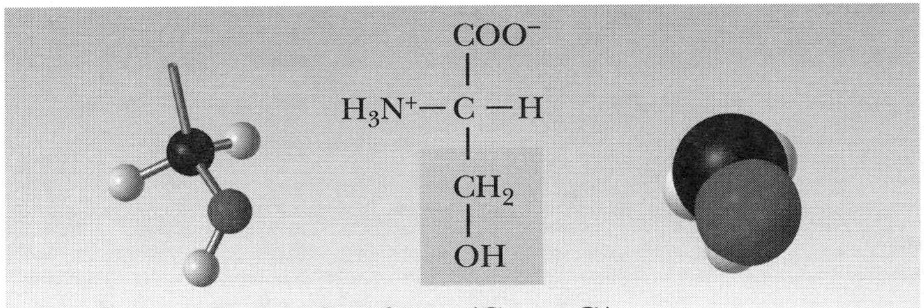

Serine (Ser, S)

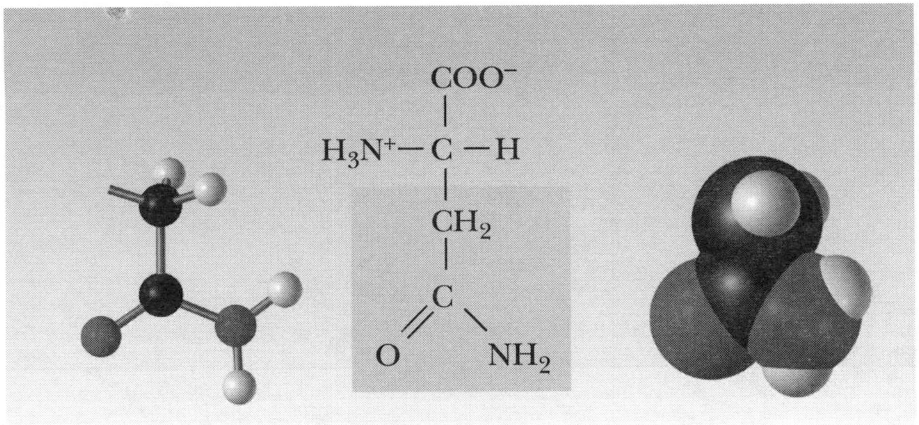

Asparagine (Asn, N)

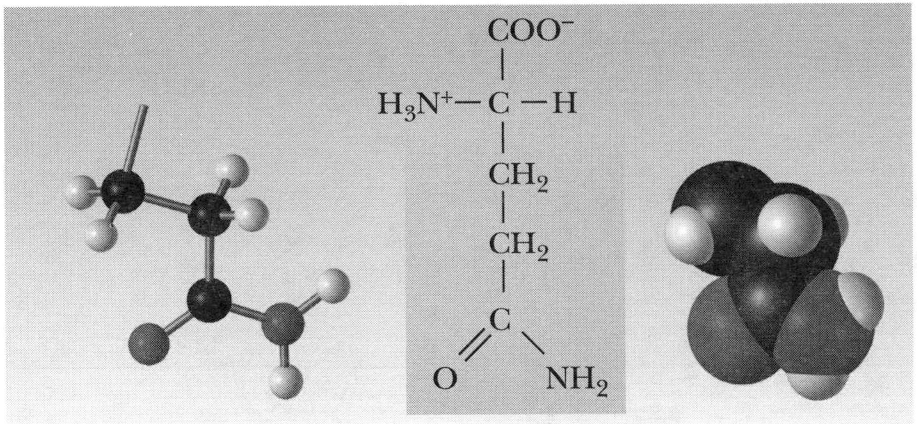

Glutamine (Gln, Q)

Figure 3.4b Amino acids and their structures

(c) Acidic

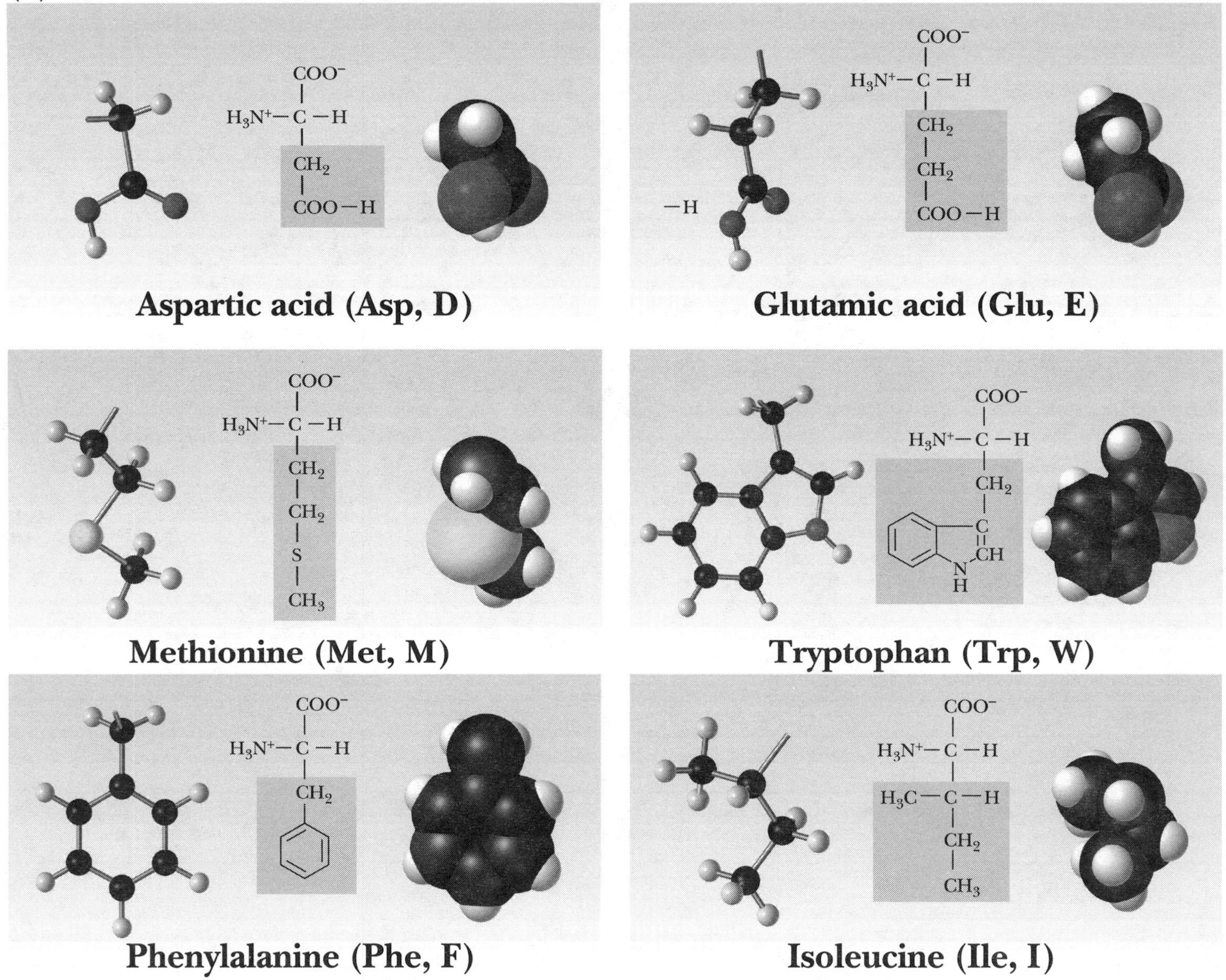

Figure 3.4c Amino acids and their structures

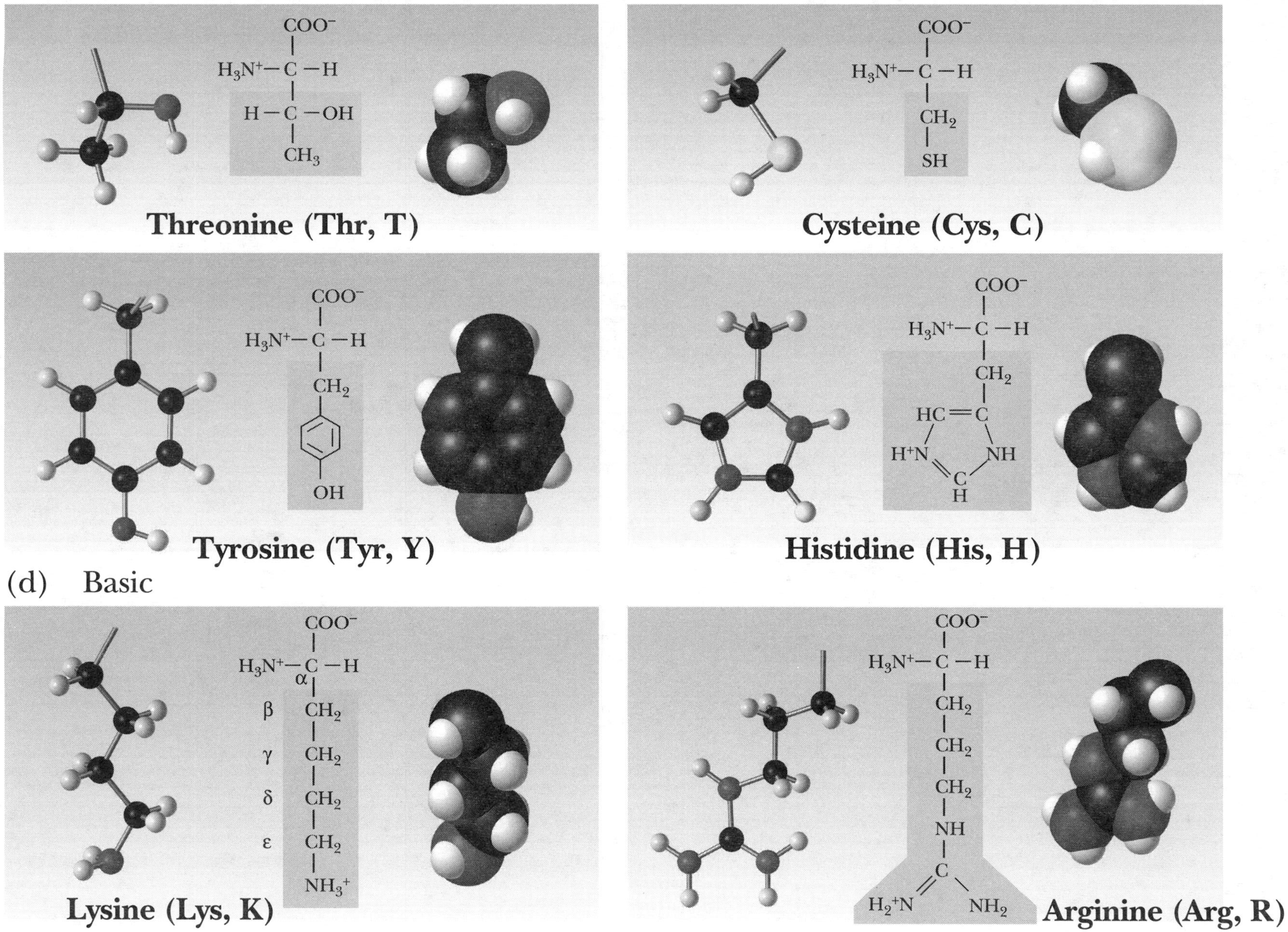

Figure 3.4c (continued), 3.4d Amino acids and their structures

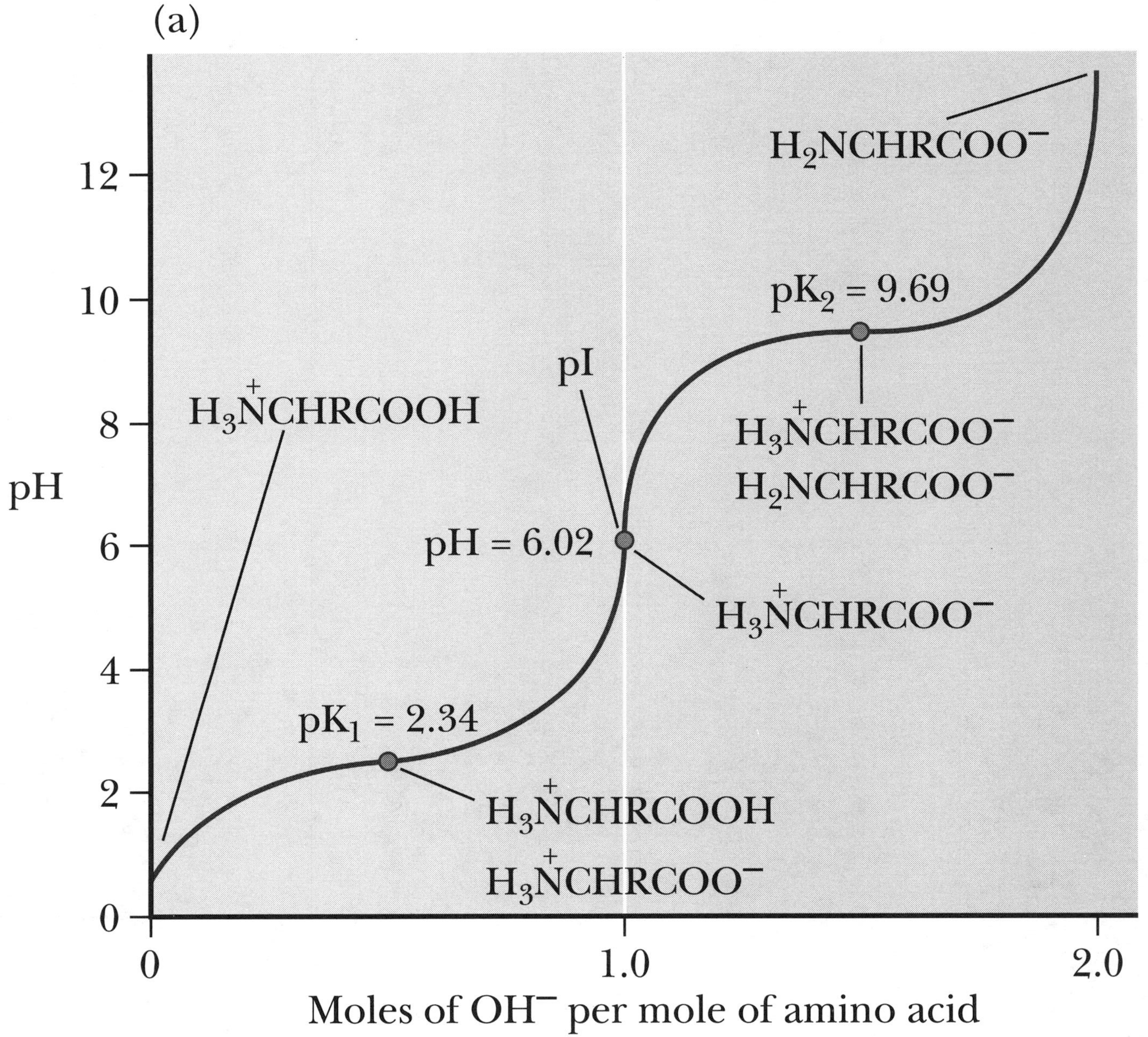
(a)
pH
12
10
8
6
4
2
0
0
1.0
2.0
Moles of OH⁻ per mole of amino acid
H₃N⁺CHRCOOH
pK₁ = 2.34
H₃N⁺CHRCOOH
H₃N⁺CHRCOO⁻
pI
pH = 6.02
H₃N⁺CHRCOO⁻
pK₂ = 9.69
H₃N⁺CHRCOO⁻
H₂NCHRCOO⁻
H₂NCHRCOO⁻

Figure 3.7a Titration curves of amino acids

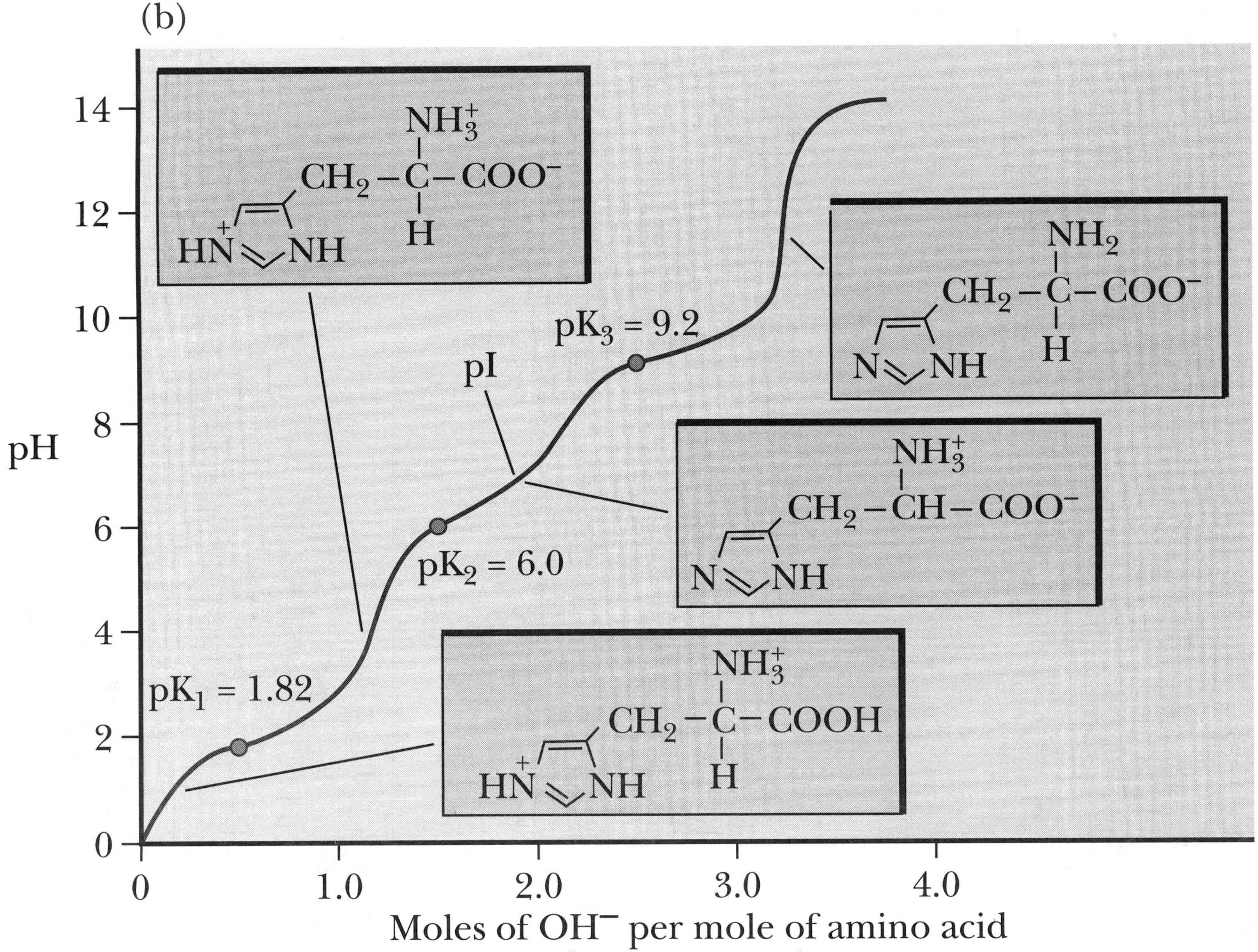
(b)
pH
14
12
10
8
6
4
2
0
1.0
2.0
3.0
4.0
Moles of OH⁻ per mole of amino acid
pK1 = 1.82
pK2 = 6.0
pK3 = 9.2
pI
NH3+
CH2–C–COO⁻
H
HN+ NH
NH2
CH2–C–COO⁻
H
N NH
NH3+
CH2–CH–COO⁻
N NH
NH3+
CH2–C–COOH
H
HN+ NH

Figure 3.7b Titration curves of amino acids

TABLE 3.1 Names and Abbreviations of the Common Amino Acids

Amino Acid	Three-Letter Abbreviation	One-Letter Abbreviation
Alanine	Ala	A
Arginine	Arg	R
Asparagine	Asn	N
Aspartic acid	Asp	D
Cysteine	Cys	C
Glutamic acid	Glu	E
Glutamine	Gln	Q
Glycine	Gly	G
Histidine	His	H
Isoleucine	Ile	I
Leucine	Leu	L
Lysine	Lys	K
Methionine	Met	M
Phenylalanine	Phe	F
Proline	Pro	P
Serine	Ser	S
Threonine	Thr	T
Tryptophan	Trp	W
Tyrosine	Tyr	Y
Valine	Val	V

Table 3.1 Common amino acids and their abbreviations

TABLE 3.2 pK_a Values of Common Amino Acids

Acid	α-COOH	α-NH_3^+	RH or RH^+
Gly	2.34	9.60	
Ala	2.34	9.69	
Val	2.32	9.62	
Leu	2.36	9.68	
Ile	2.36	9.68	
Ser	2.21	9.15	
Thr	2.63	10.43	
Met	2.28	9.21	
Phe	1.83	9.13	
Trp	2.38	9.39	
Asn	2.02	8.80	
Gln	2.17	9.13	
Pro	1.99	10.6	
Asp	2.09	9.82	3.86*
Glu	2.19	9.67	4.25*
His	1.82	9.17	6.0*
Cys	1.71	10.78	8.33*
Tyr	2.20	9.11	10.07
Lys	2.18	8.95	10.53
Arg	2.17	9.04	12.48

*For these amino acids the R group ionization occurs before the α-NH_3^+ ionization.

Table 3.2 PKa values of common amino acids

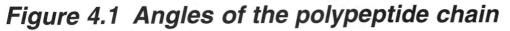

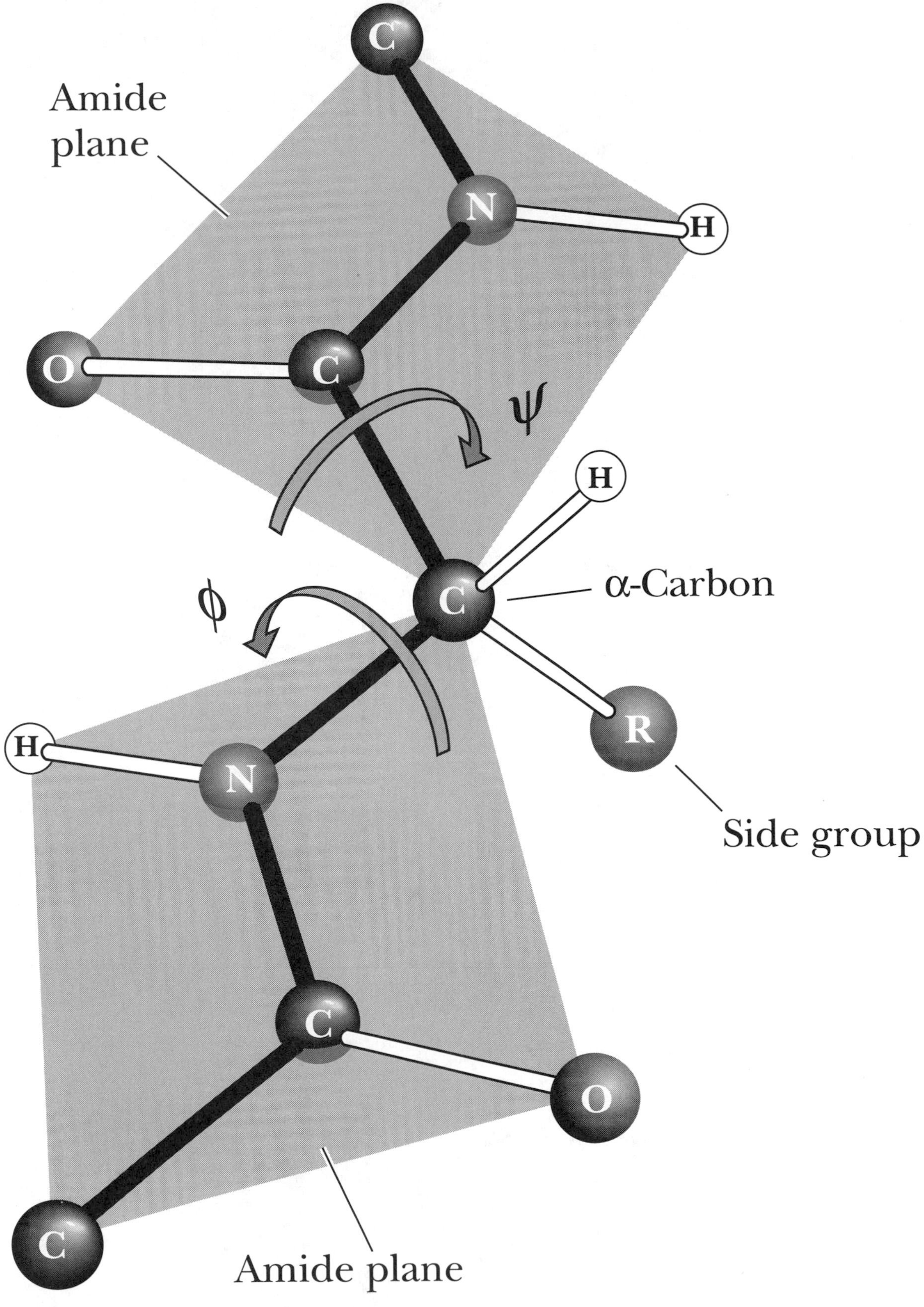

$\phi = 180°, \psi = 180°$

Figure 4.1 Angles of the polypeptide chain

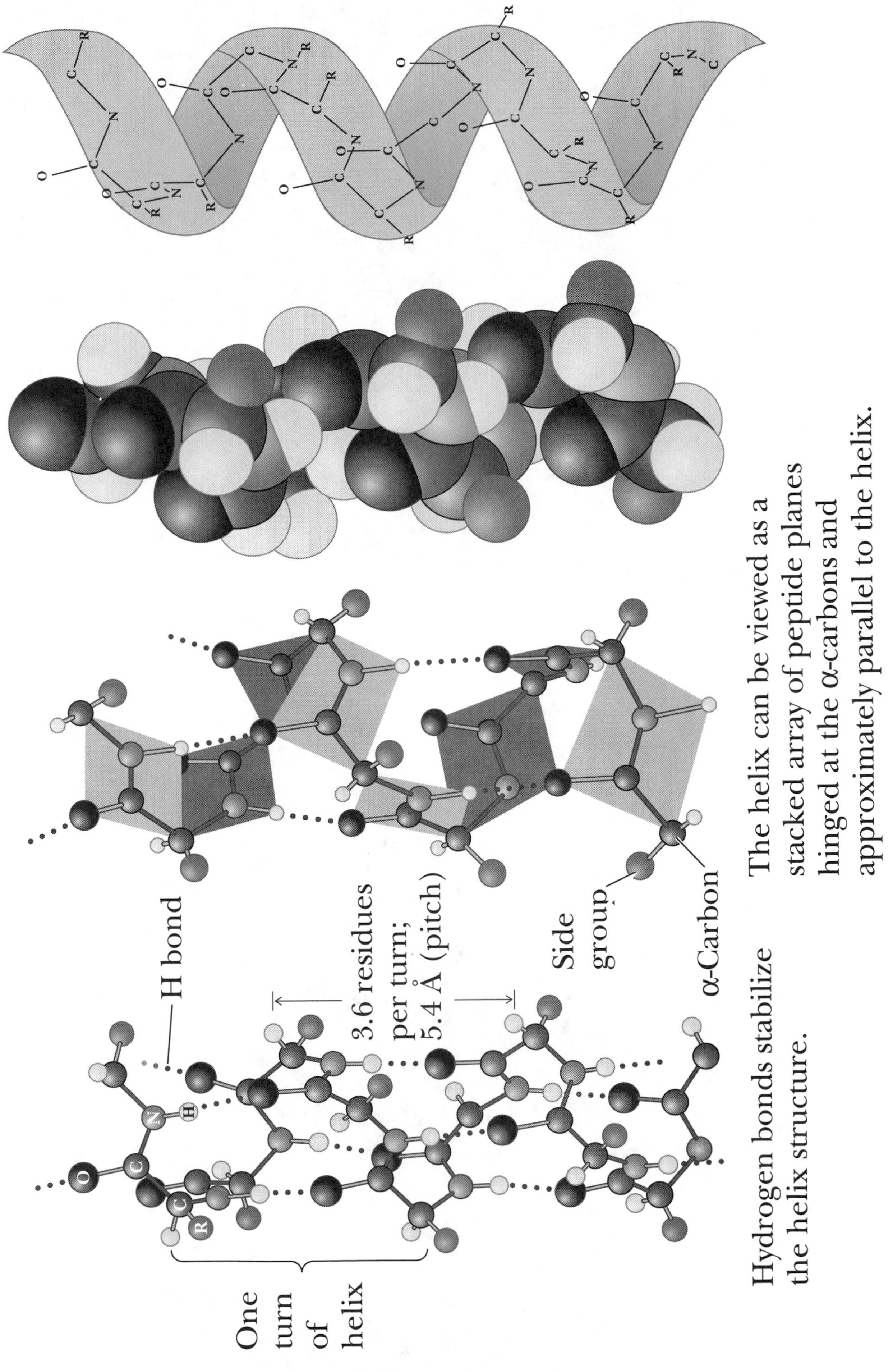

One turn of helix
H bond
3.6 residues per turn; 5.4 Å (pitch)
Side group
α-Carbon
Hydrogen bonds stabilize the helix structure.
The helix can be viewed as a stacked array of peptide planes hinged at the α-carbons and approximately parallel to the helix.

Figure 4.2 The α-helix

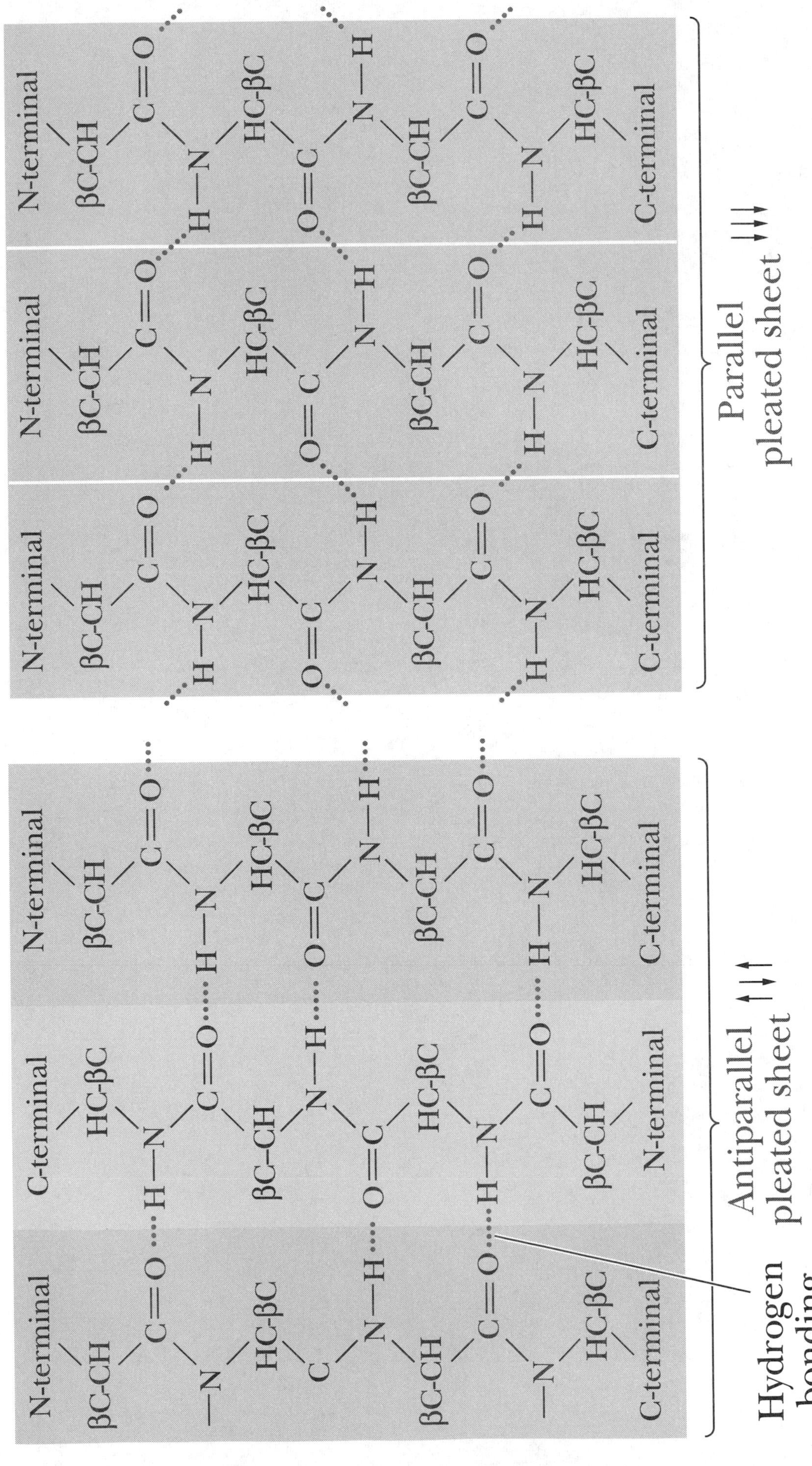
N-terminal
C-terminal
Hydrogen bonding
Antiparallel pleated sheet
Parallel pleated sheet

Figure 4.3 β-pleated sheet structures

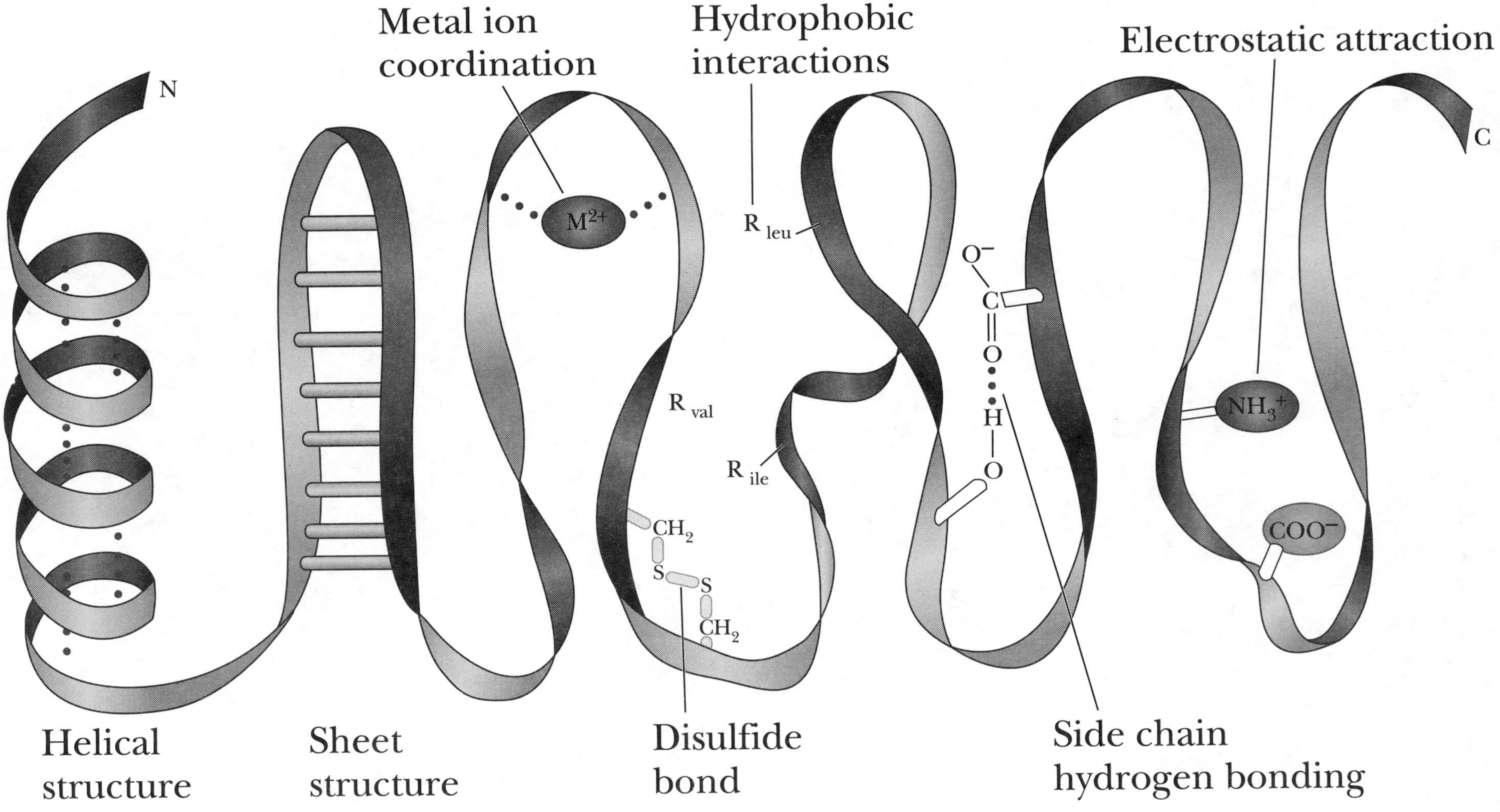

Figure 4.12 Forces that stabilize the tertiary structure of proteins

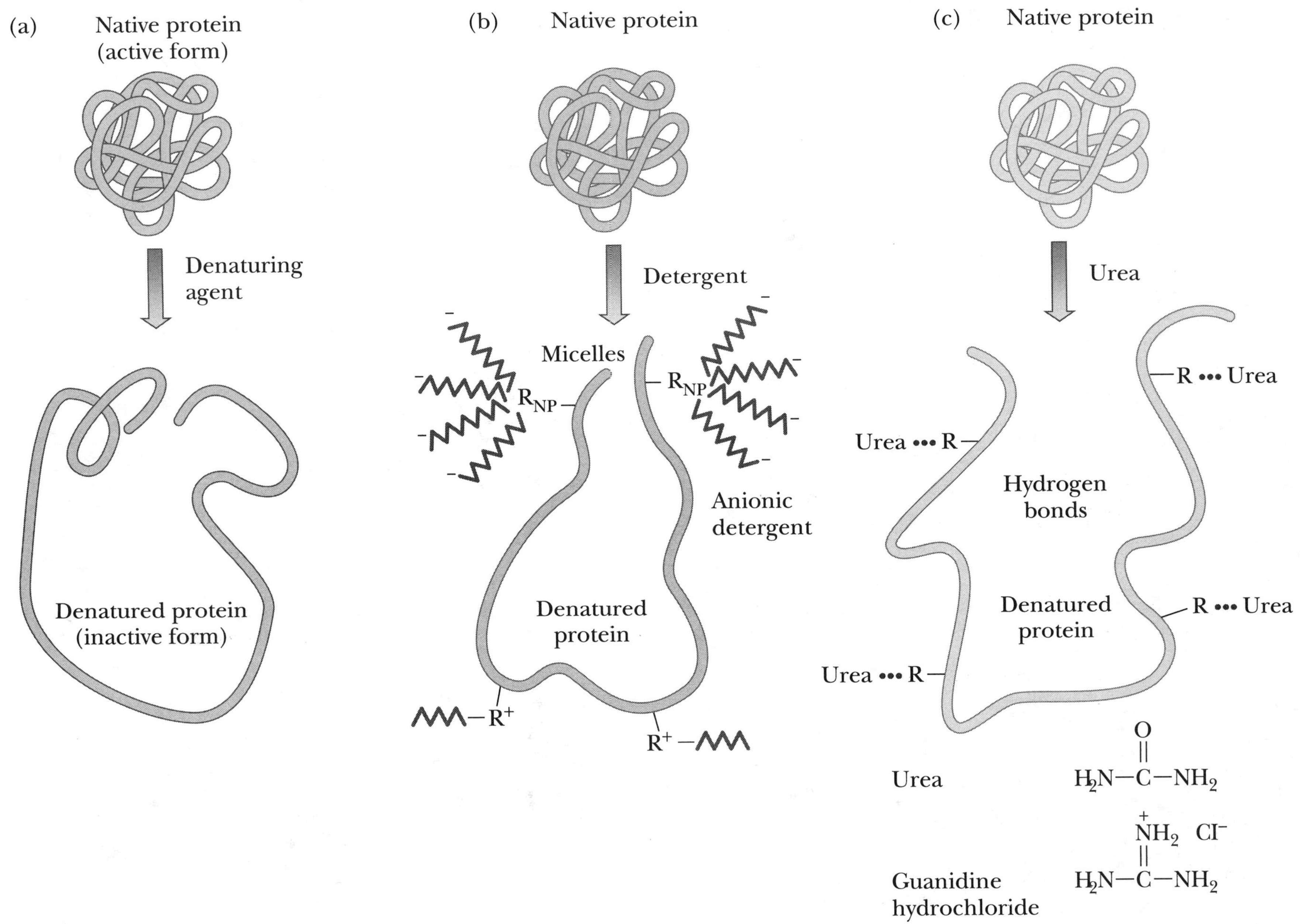
(a)
Native protein (active form)
Denaturing agent
Denatured protein (inactive form)
(b)
Native protein
Detergent
Micelles
R_{NP}
Anionic detergent
Denatured protein
R^+
(c)
Native protein
Urea
R ••• Urea
Urea ••• R
Hydrogen bonds
Denatured protein
Urea
$H_2N-C(=O)-NH_2$
Guanidine hydrochloride
$H_2N-C(=\overset{+}{N}H_2)-NH_2\ Cl^-$

Figure 4.18 Denaturation of proteins

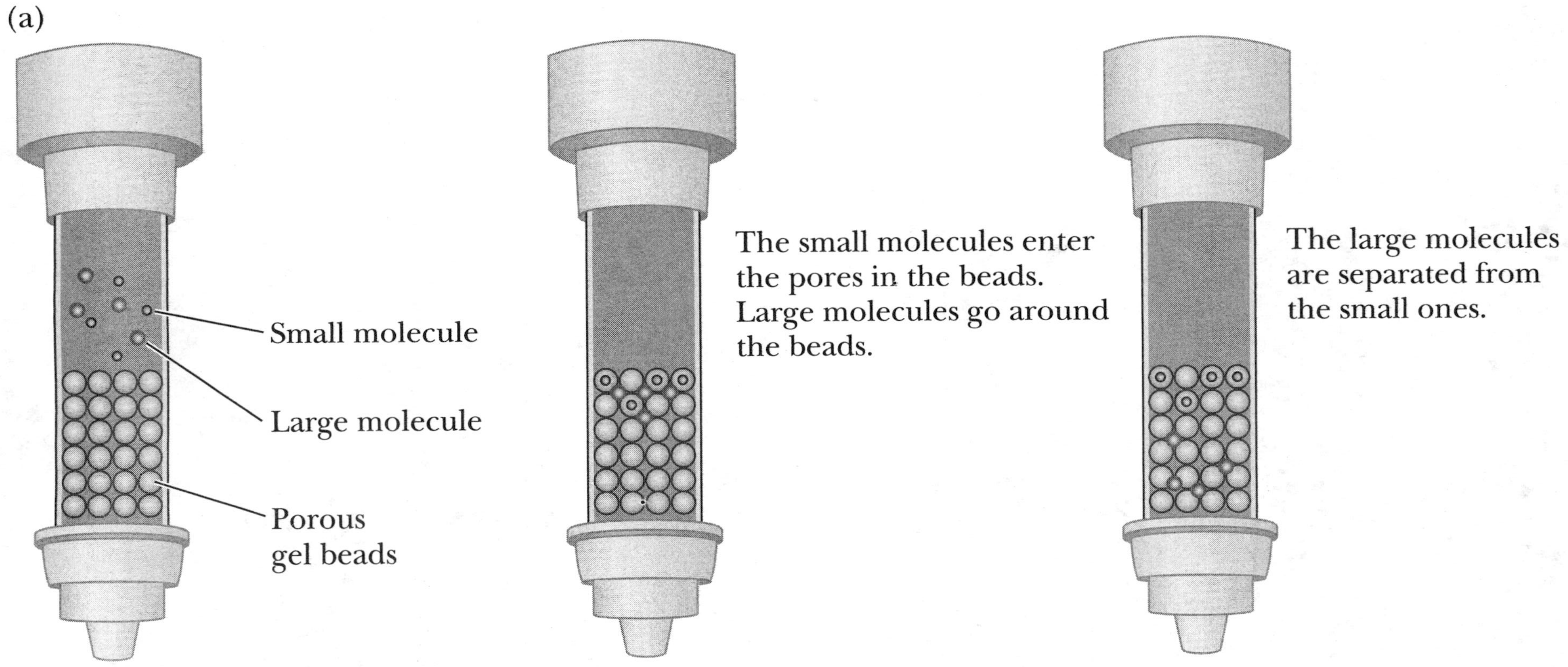
(a)
Small molecule
Large molecule
Porous
gel beads
The small molecules enter
the pores in the beads.
Large molecules go around
the beads.
The large molecules
are separated from
the small ones.

Figure A.5a Gel filtration chromatography

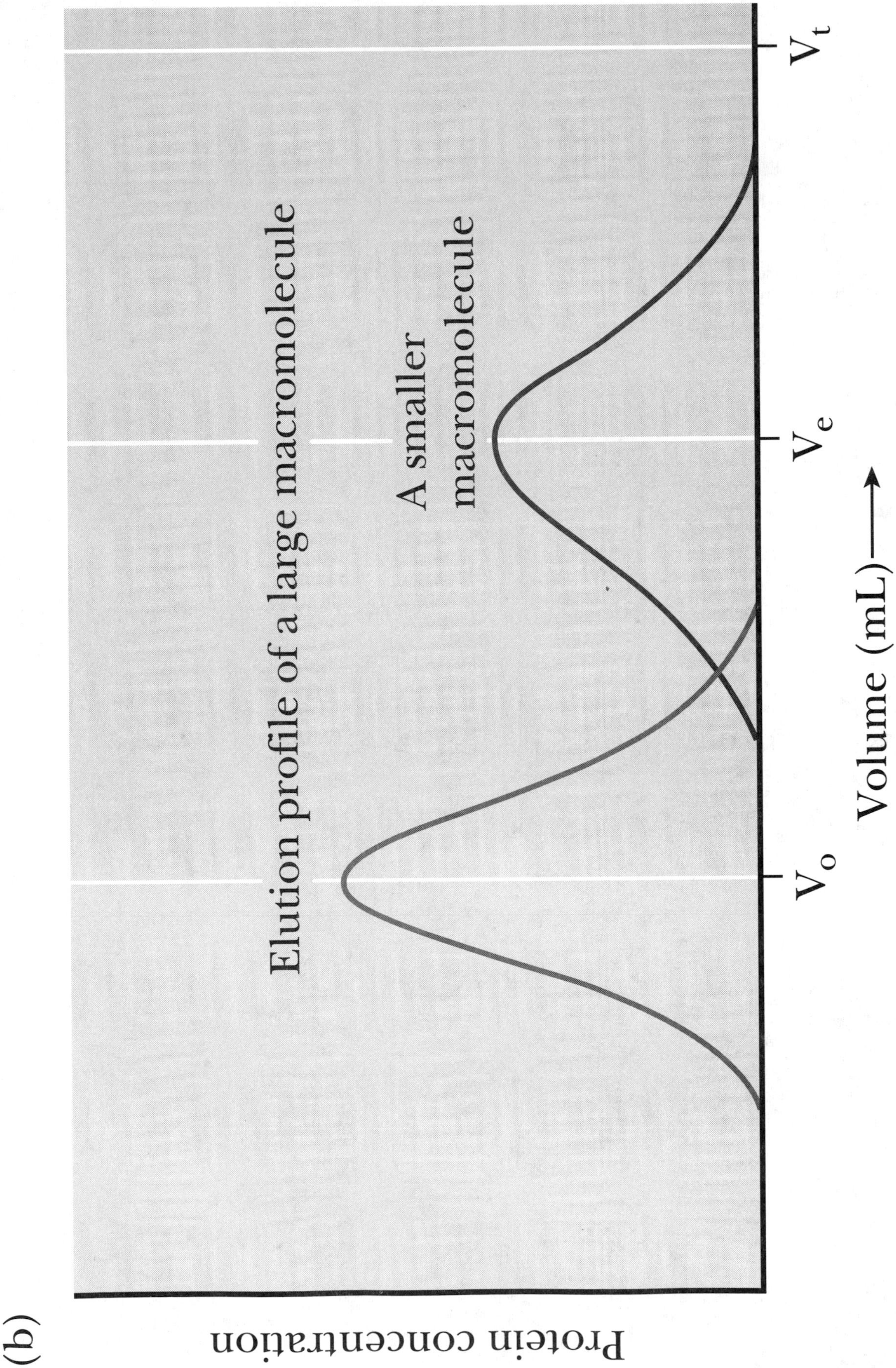

Figure A.5b Gel filtration chromatography

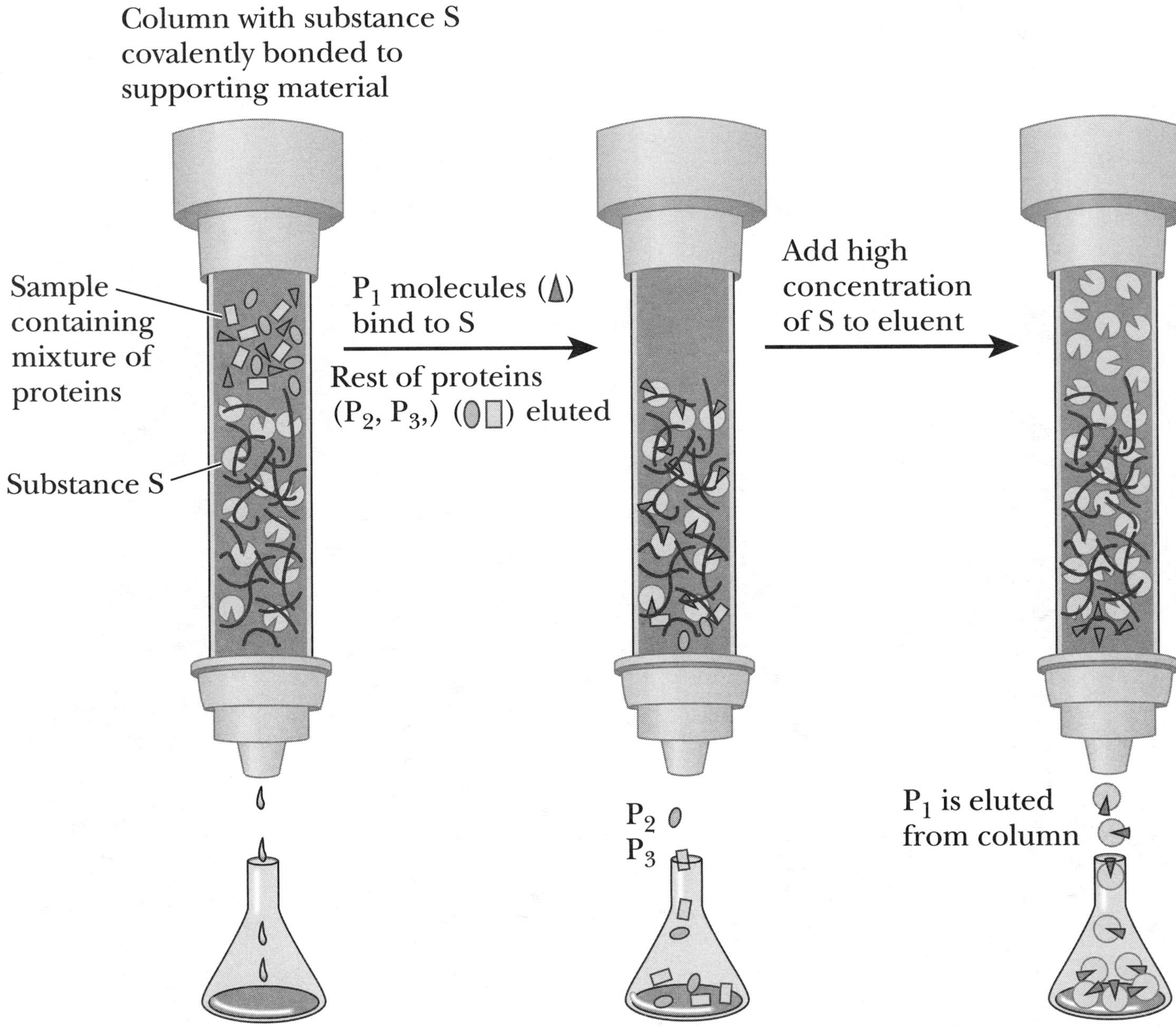

Column with substance S covalently bonded to supporting material
Sample containing mixture of proteins
Substance S
P_1 molecules (▲) bind to S
Rest of proteins (P_2, P_3,) (⬮▭) eluted
Add high concentration of S to eluent
P_2
P_3
P_1 is eluted from column

Figure A.6 Affinity chromatography

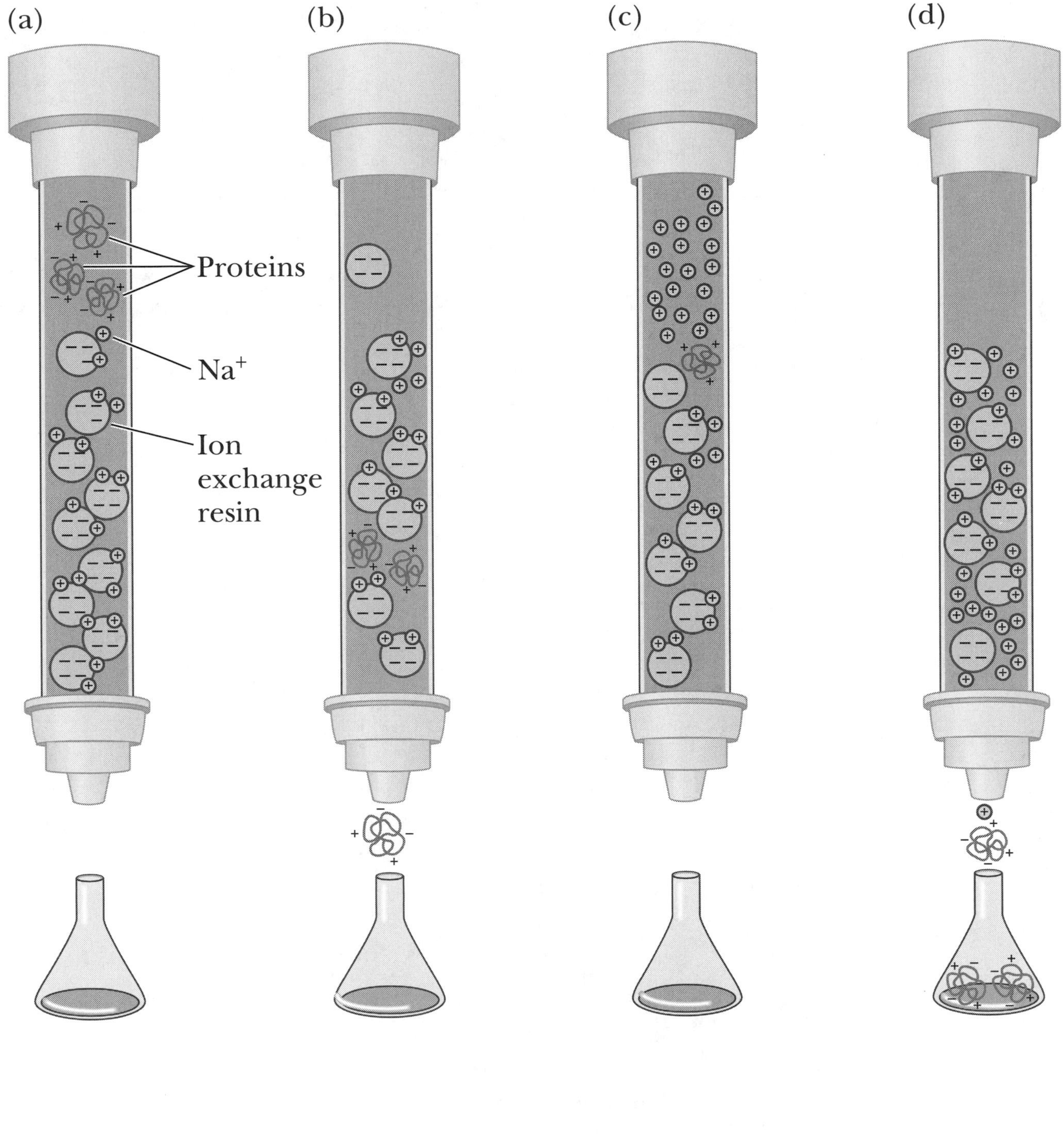

Figure A.8 Ion exchange chromatography using a cation exchanger

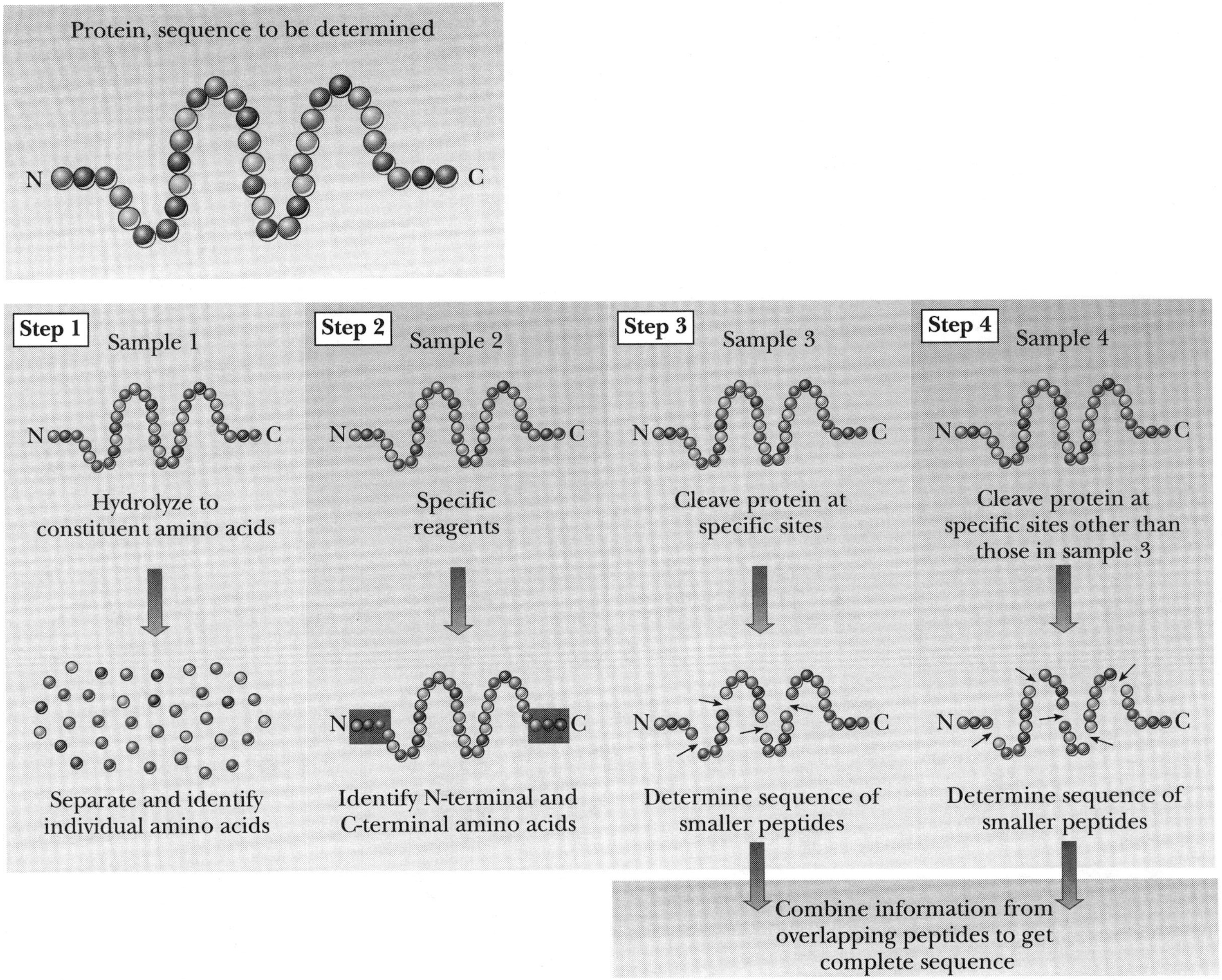

Protein, sequence to be determined
N
C
Step 1
Sample 1
Hydrolyze to constituent amino acids
Separate and identify individual amino acids
Step 2
Sample 2
Specific reagents
Identify N-terminal and C-terminal amino acids
Step 3
Sample 3
Cleave protein at specific sites
Determine sequence of smaller peptides
Step 4
Sample 4
Cleave protein at specific sites other than those in sample 3
Determine sequence of smaller peptides
Combine information from overlapping peptides to get complete sequence

Figure A.13 The strategy for determining the primary structure of a given protein

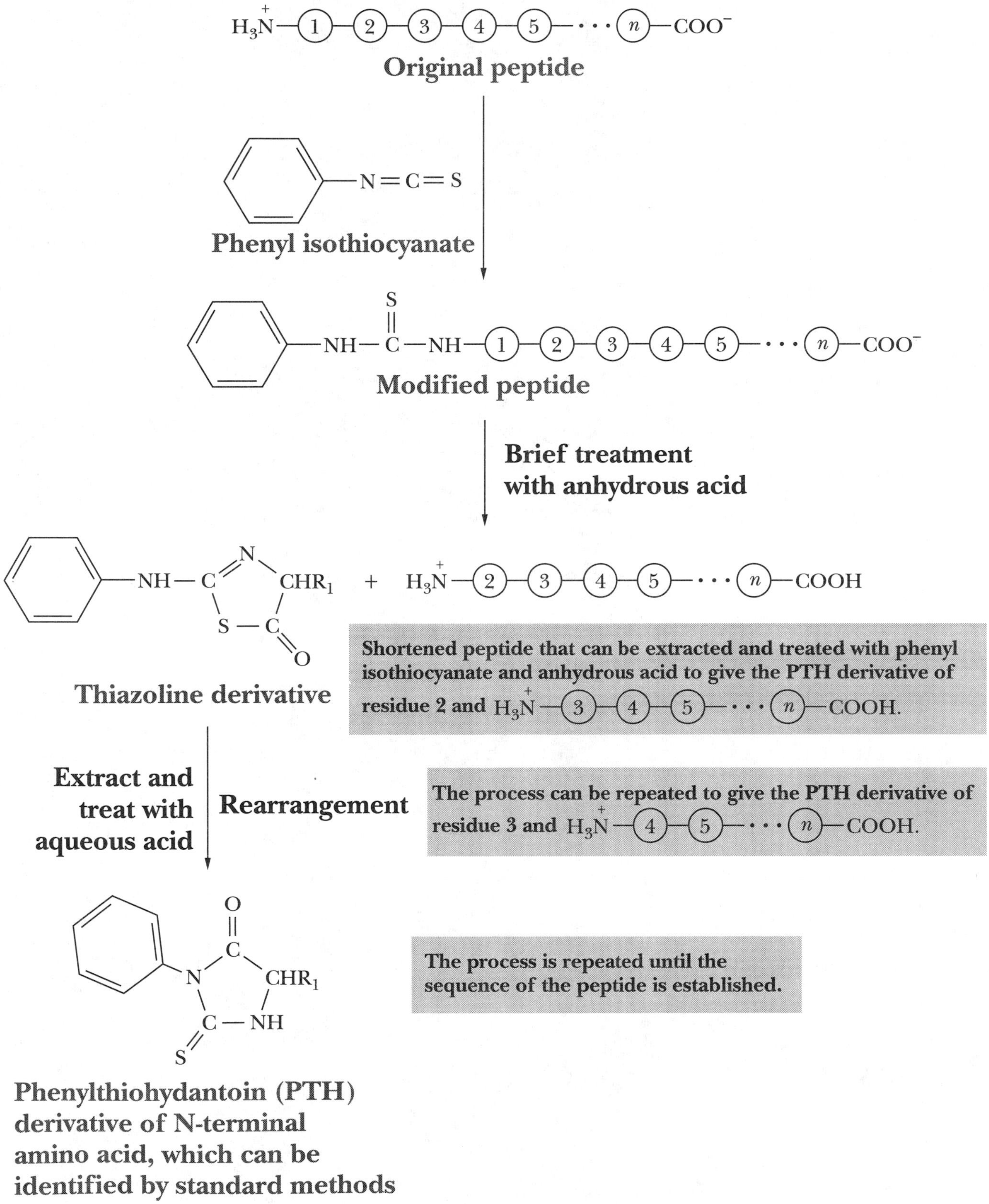
Original peptide
Phenyl isothiocyanate
Modified peptide
Brief treatment with anhydrous acid
Thiazoline derivative
Shortened peptide that can be extracted and treated with phenyl isothiocyanate and anhydrous acid to give the PTH derivative of residue 2 and H3N–3–4–5···n–COOH.
Extract and treat with aqueous acid
Rearrangement
The process can be repeated to give the PTH derivative of residue 3 and H3N–4–5···n–COOH.
The process is repeated until the sequence of the peptide is established.
Phenylthiohydantoin (PTH) derivative of N-terminal amino acid, which can be identified by standard methods

Figure A.18 Sequencing of peptides by the Edman method

(a)
Transition state
$\Delta G^{\circ\ddagger}$ = Activation energy
Free energy
Reactants
ΔG° = Free energy change
Products
Progress of reaction
(b)
Uncatalyzed
Catalyzed
Free energy
Reactants
Products
Progress of reaction

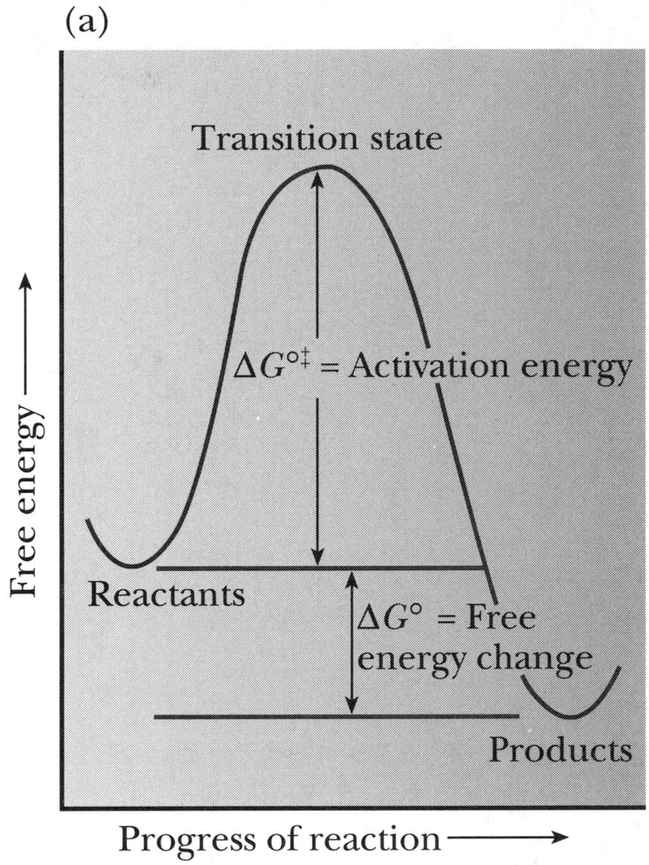

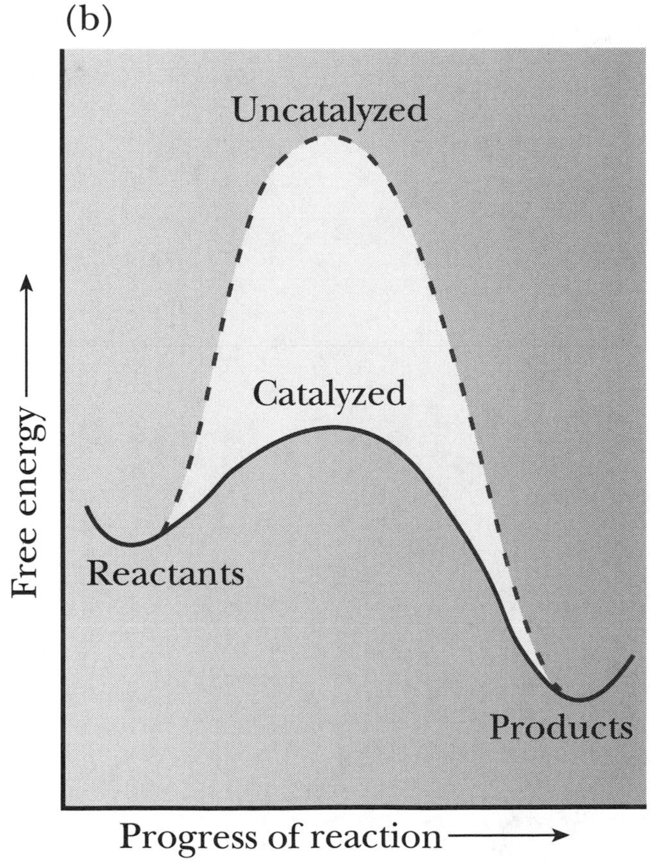

Figure 5.1 Activation energy profiles

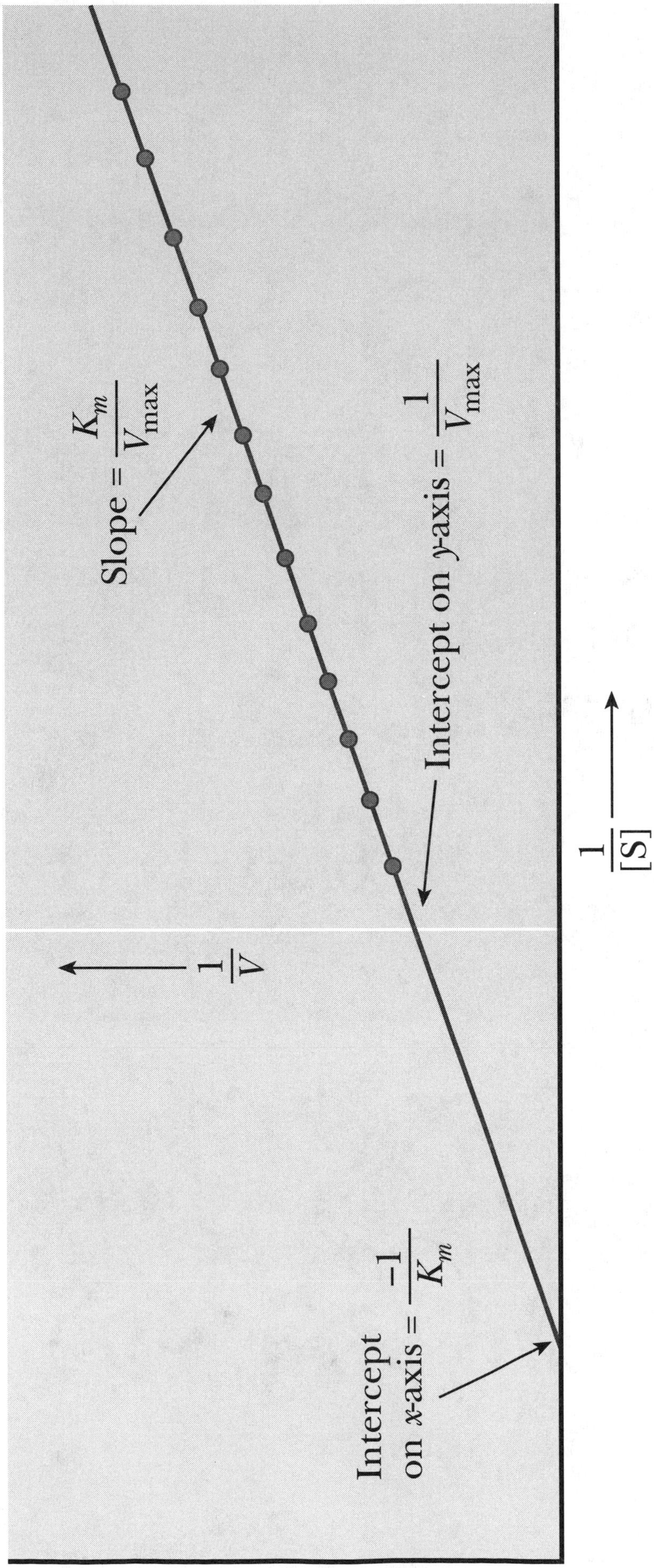

Figure 5.10 Lineweaver-Burk double-reciprocal plot of enzyme kinetics

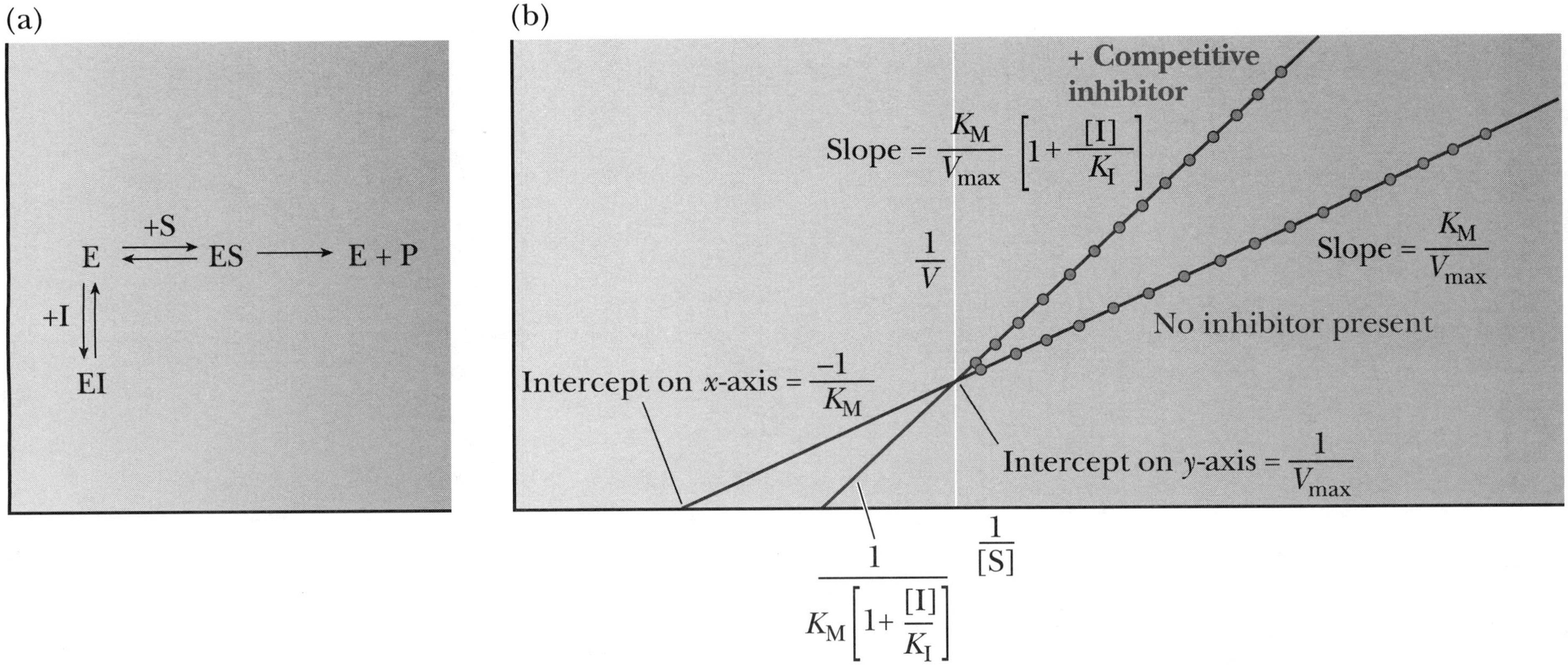

Figure 5.12 Lineweaver-Burk double reciprocal plot for competitive inhibition

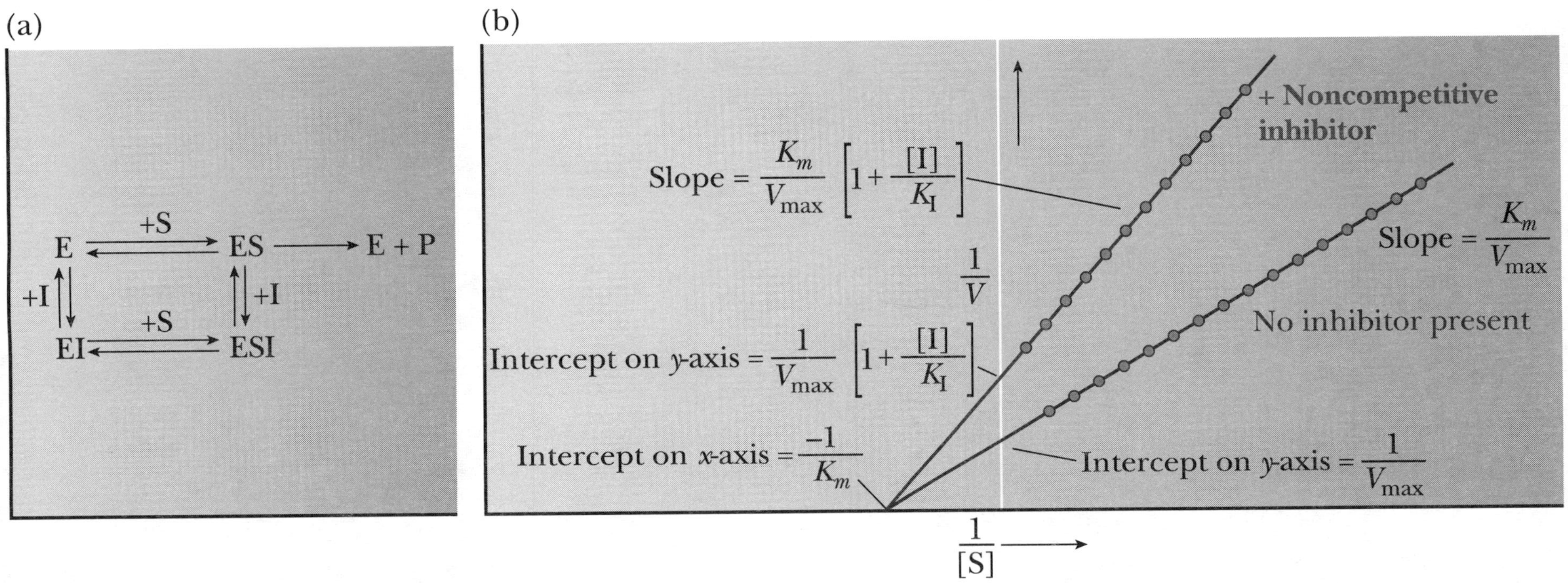

Figure 5.13 Lineweaver-Burk double reciprocal plot for noncompetitive inhibition

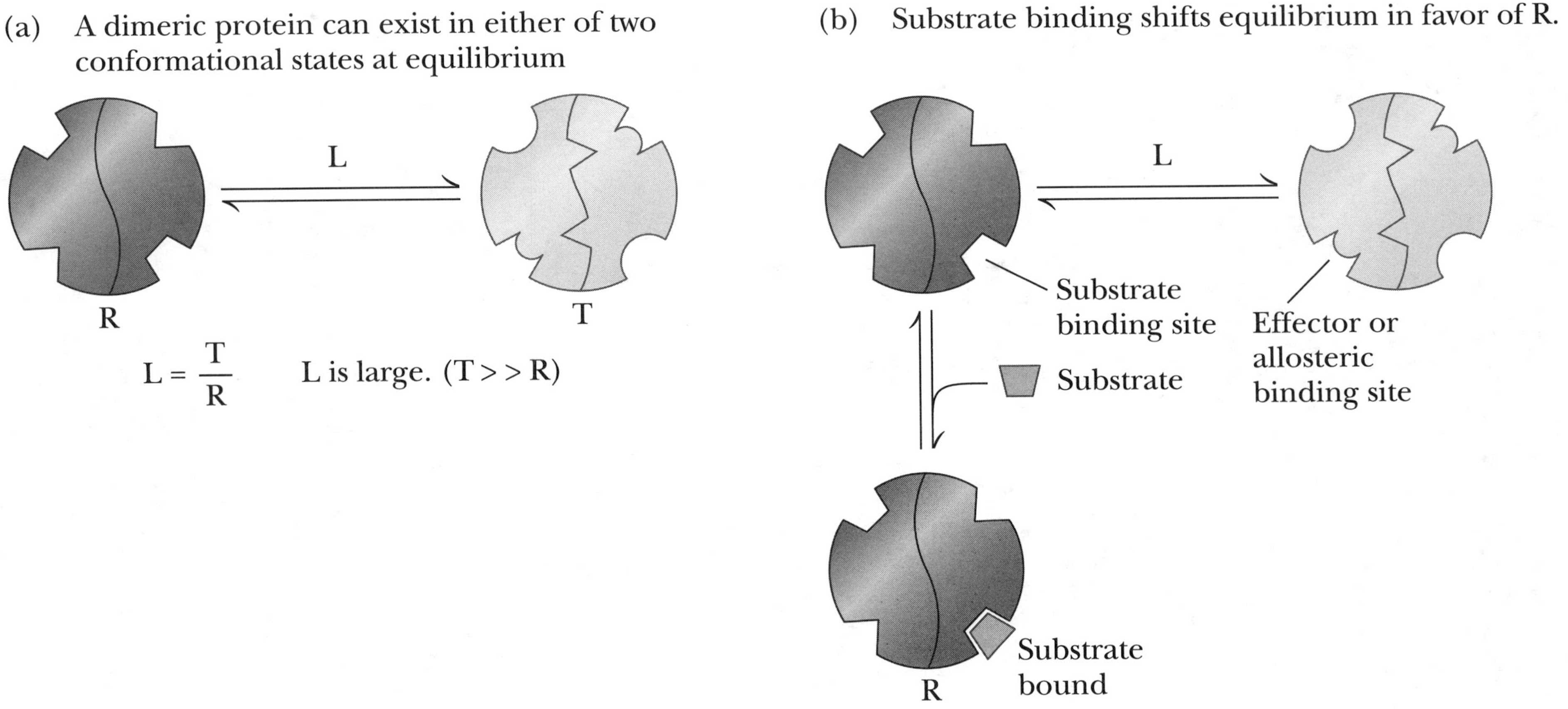

(a) A dimeric protein can exist in either of two conformational states at equilibrium
L
R
T
L = T/R
L is large. (T > > R)
(b) Substrate binding shifts equilibrium in favor of R.
L
Substrate binding site
Substrate
Effector or allosteric binding site
Substrate bound
R

Figure 6.4 Monod-Wyman-Changeux model

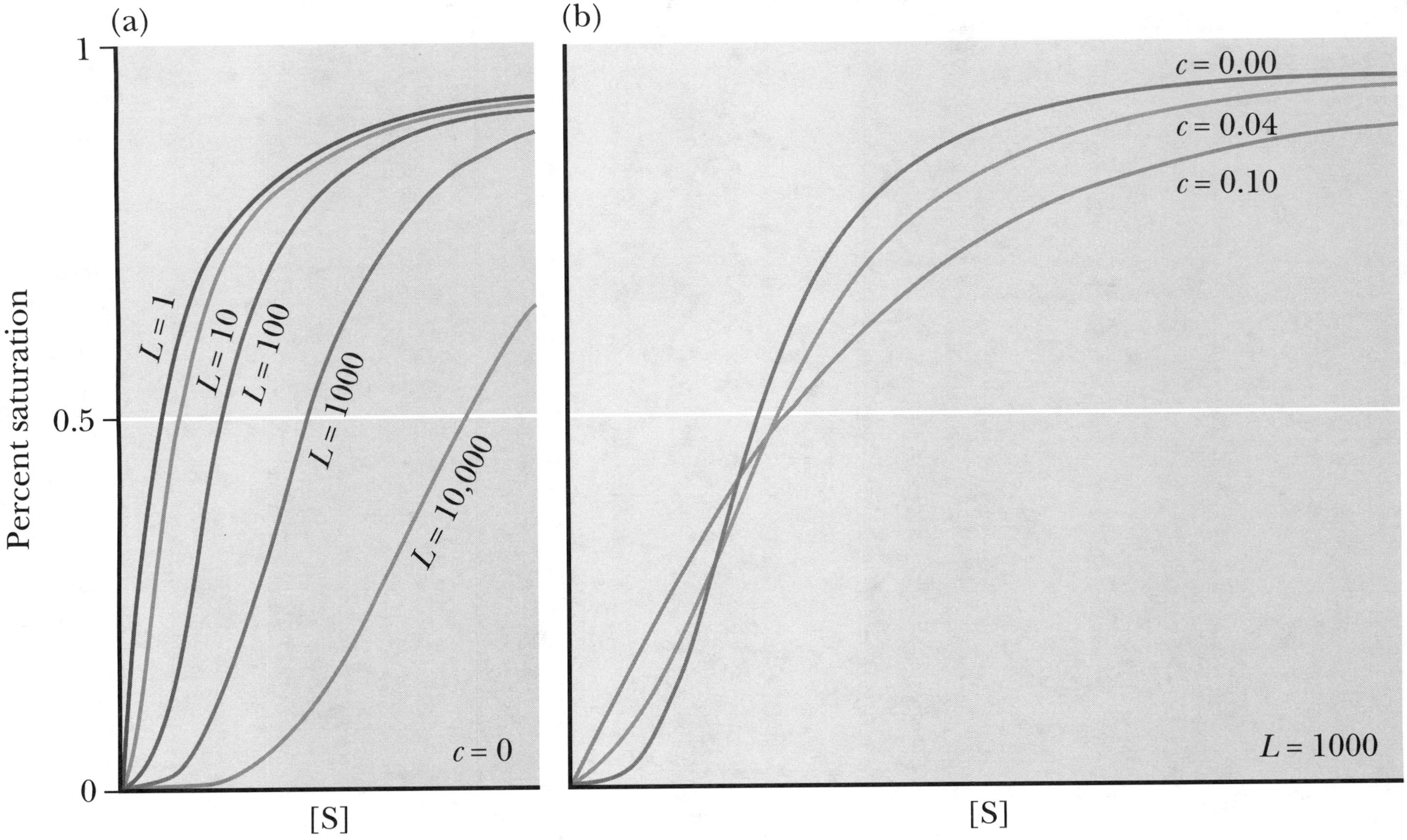

Figure 6.5 Monod-Wyman-Changeux or concerted model

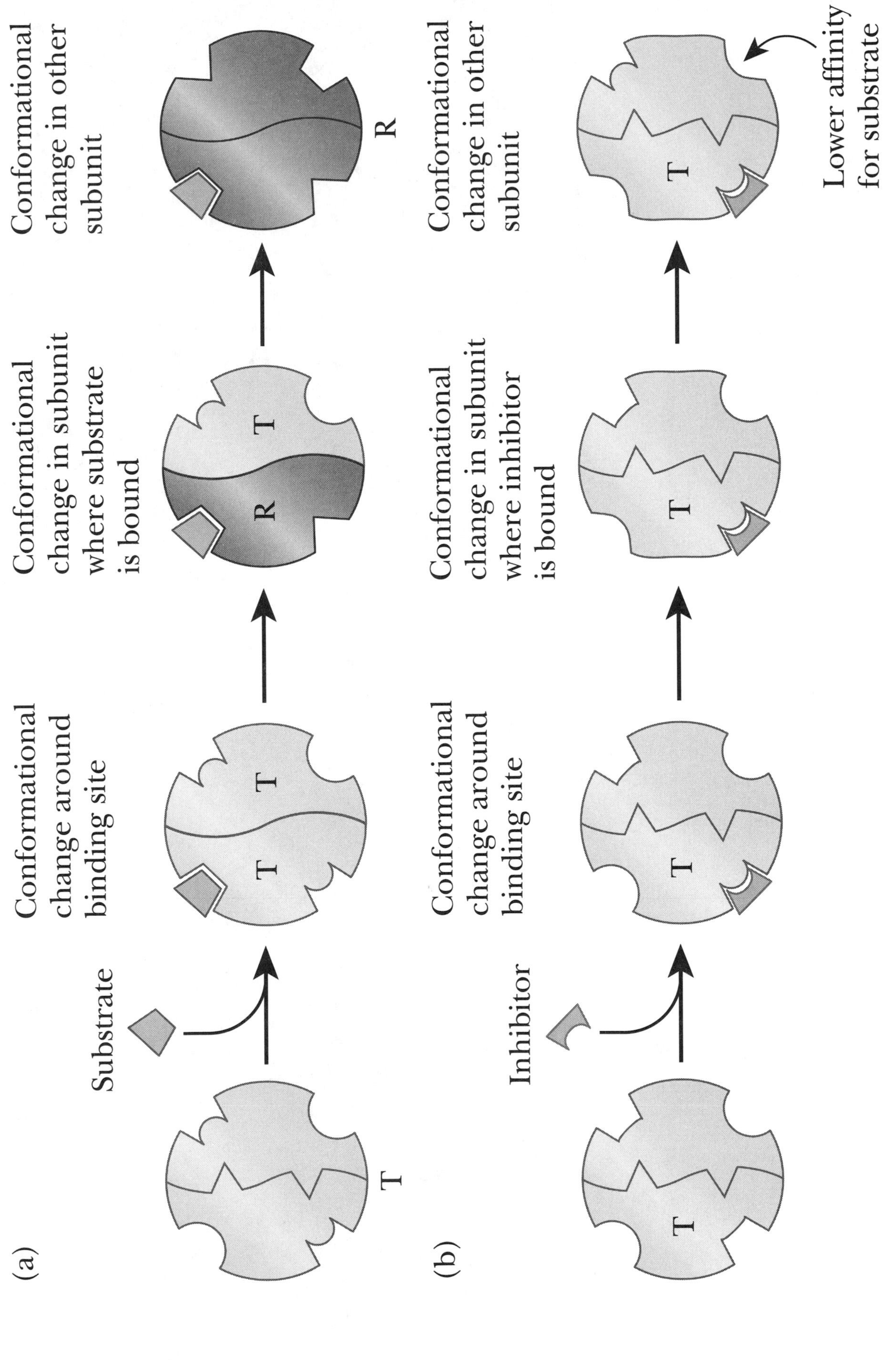

Figure 6.7 Sequential model of cooperative binding of substrates to allosteric enzyme

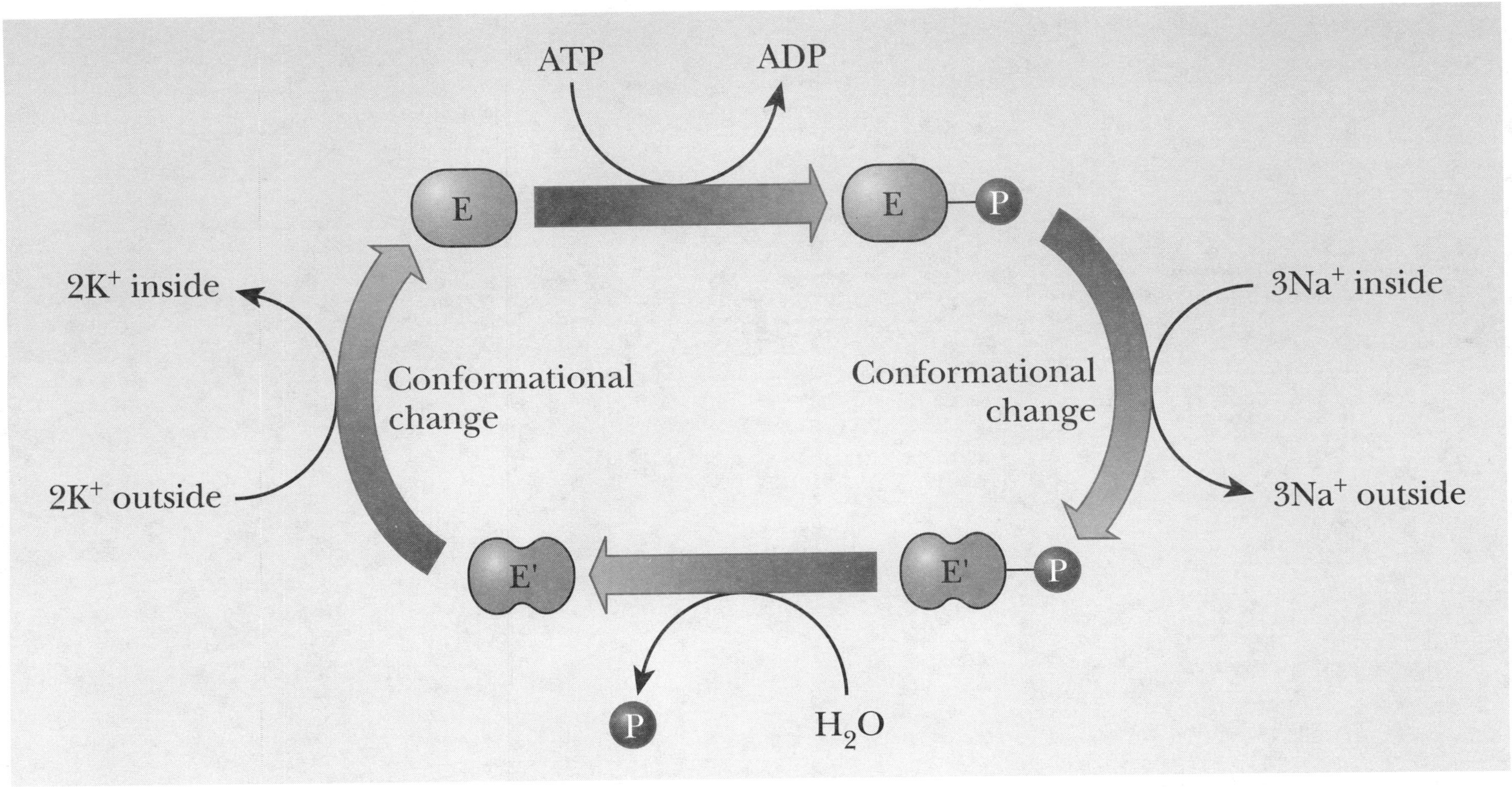

Figure 6.8 Phosphorylation of the sodium/potassium pump

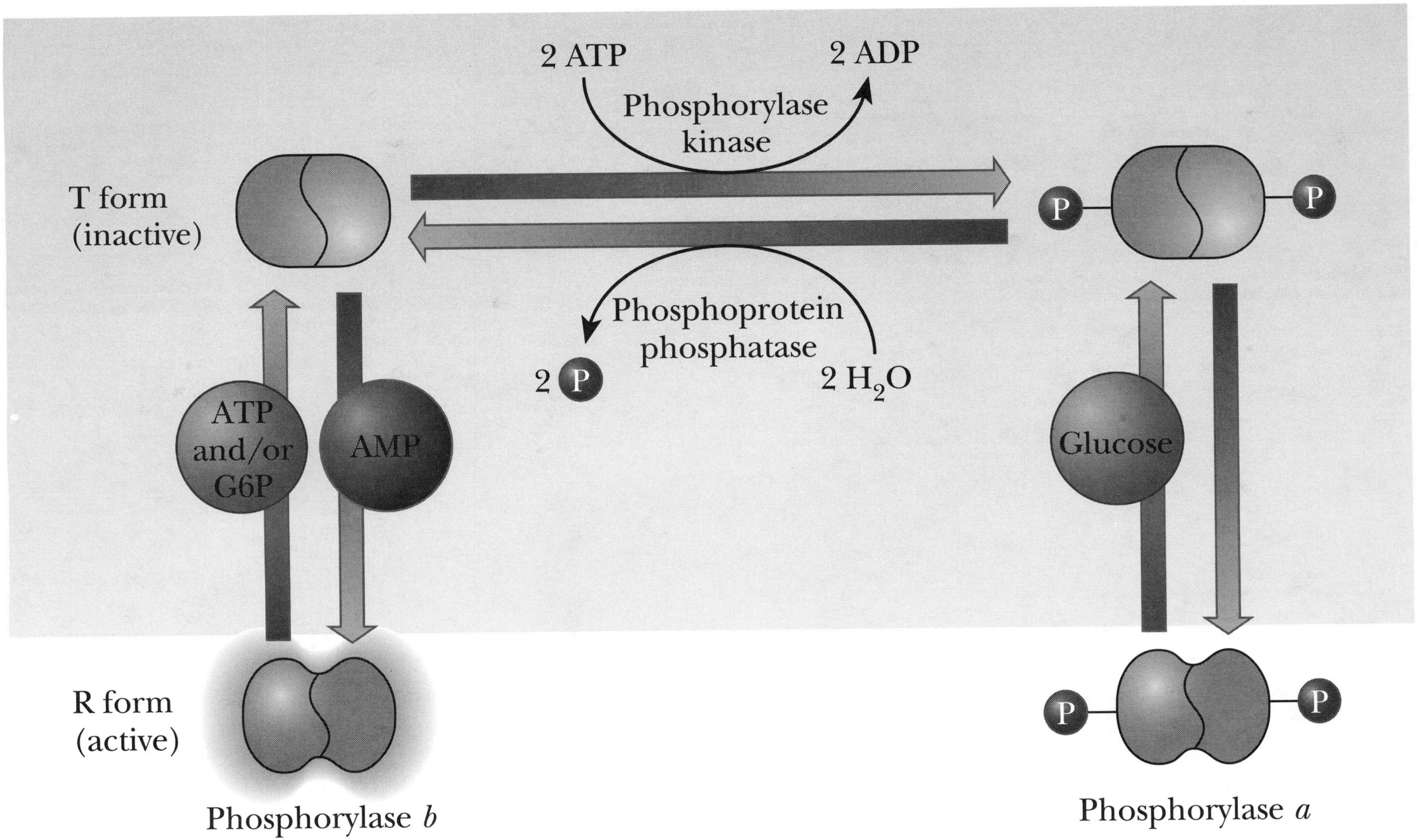

2 ATP
2 ADP
Phosphorylase
kinase
T form
(inactive)
P
P
Phosphoprotein
phosphatase
2 P
2 H_2O
ATP
and/or
G6P
AMP
Glucose
R form
(active)
P
P
Phosphorylase b
Phosphorylase a

Figure 6.9 Glycogen phosphorylation activity

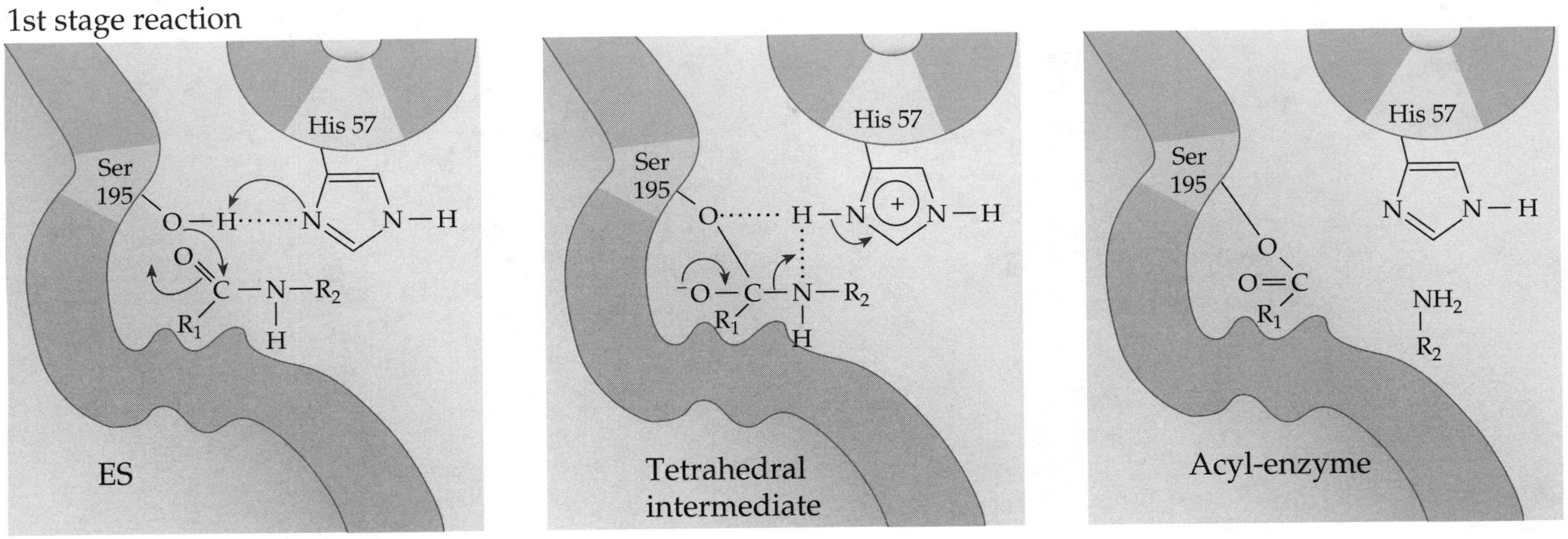

Figure 6.13 (top) The mechanism of chymotrypsin action

2nd stage reaction

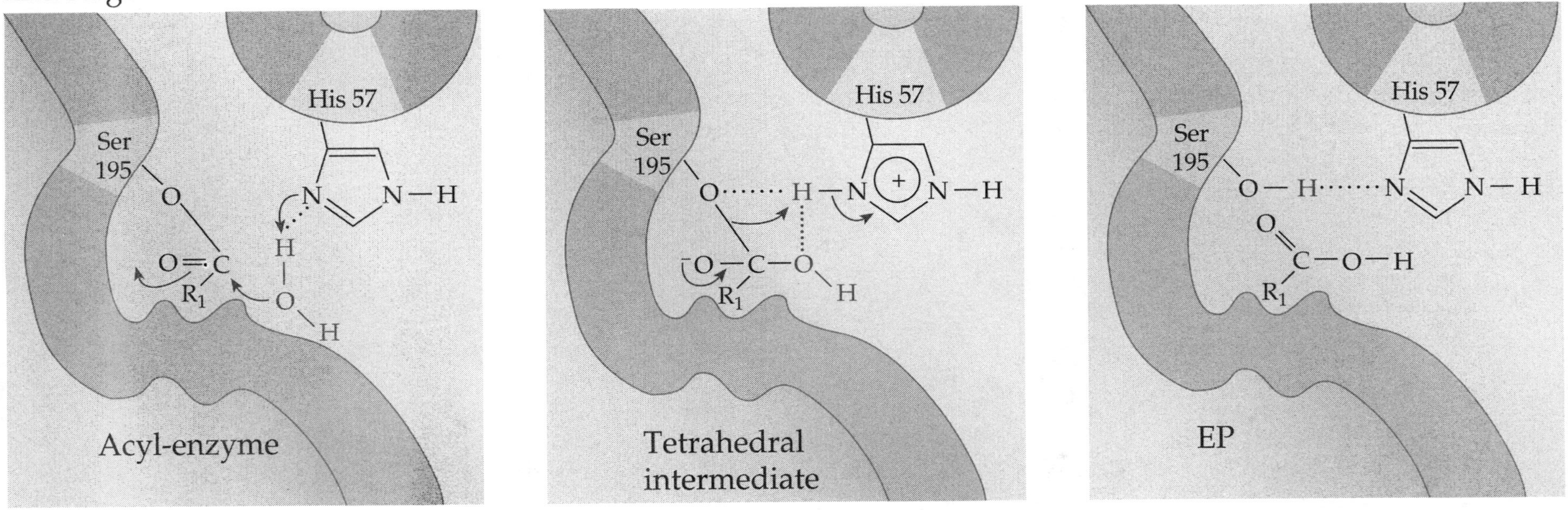

Figure 6.13 (bottom) The mechanism of chymotrypsin action

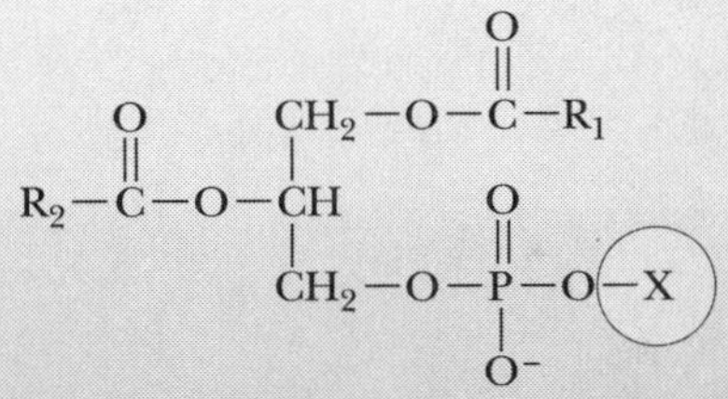

X— OH
is a second
alcohol esterified
to phosphoric acid

Name of X— OH	Formula of — X	3D structure of — X	Name of Phosphoacylglycerol
Ethanolamine	$-CH_2CH_2NH_3^+$	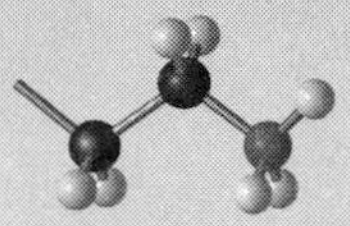	Phosphatidylethanolamine
Choline	$-CH_2CH_2\overset{+}{N}(CH_3)_3$	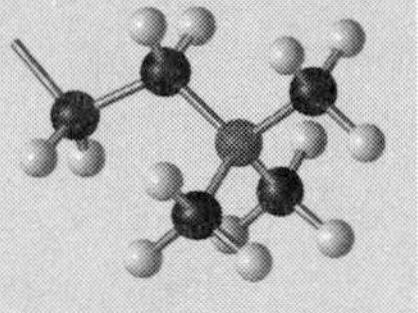	Phosphatidylcholine (lecithin)
Serine	$-CH_2CH(\overset{+}{N}H_3)COO^-$	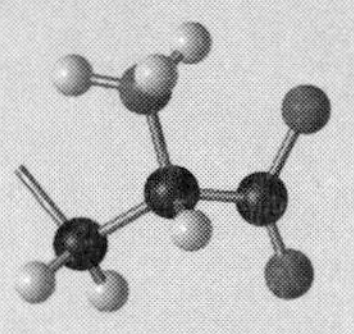	Phosphatidylserine
Inositol	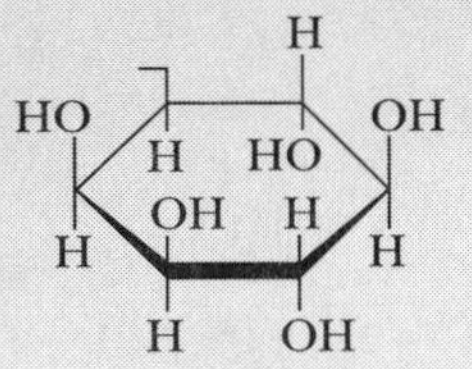	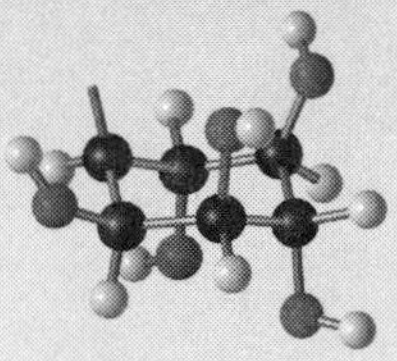	Phosphatidylinositol
Glycerol	$-CH_2CH(OH)CH_2OH$	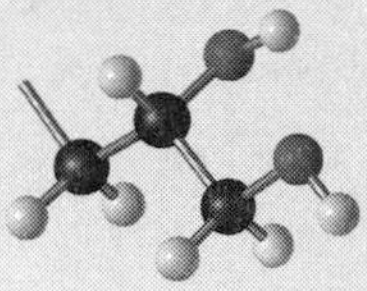	Phosphatidylglycerol
Phosphatidylglycerol	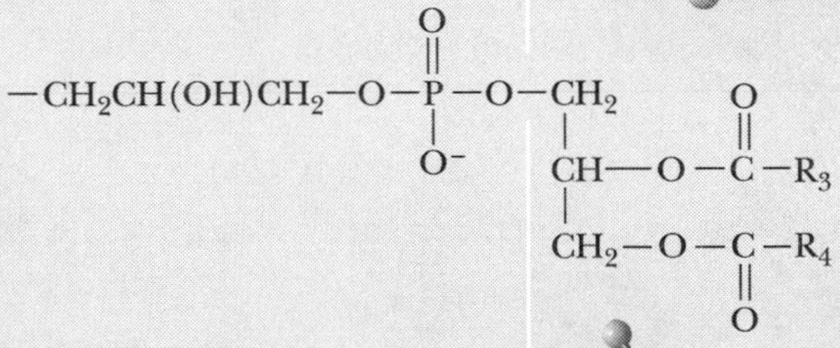	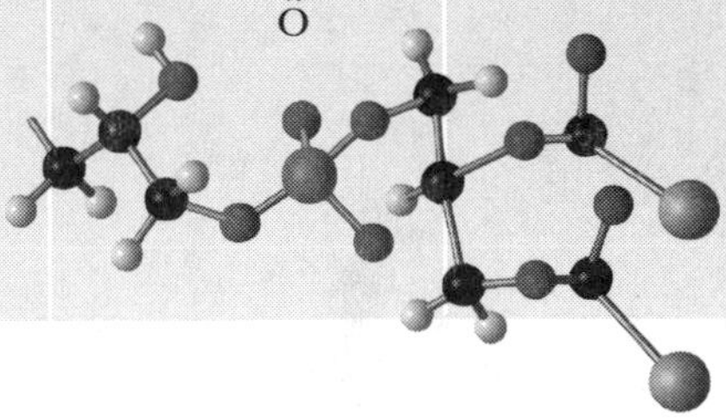	Diphosphatidylglycerol (cardiolipin)

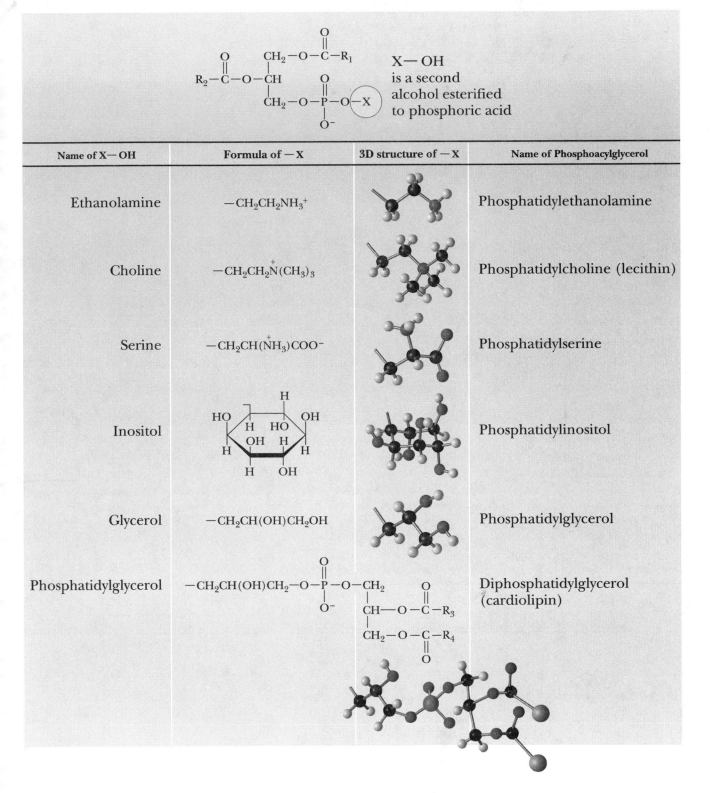

Figure 7.5 Structures of some phosphoacylglycerols

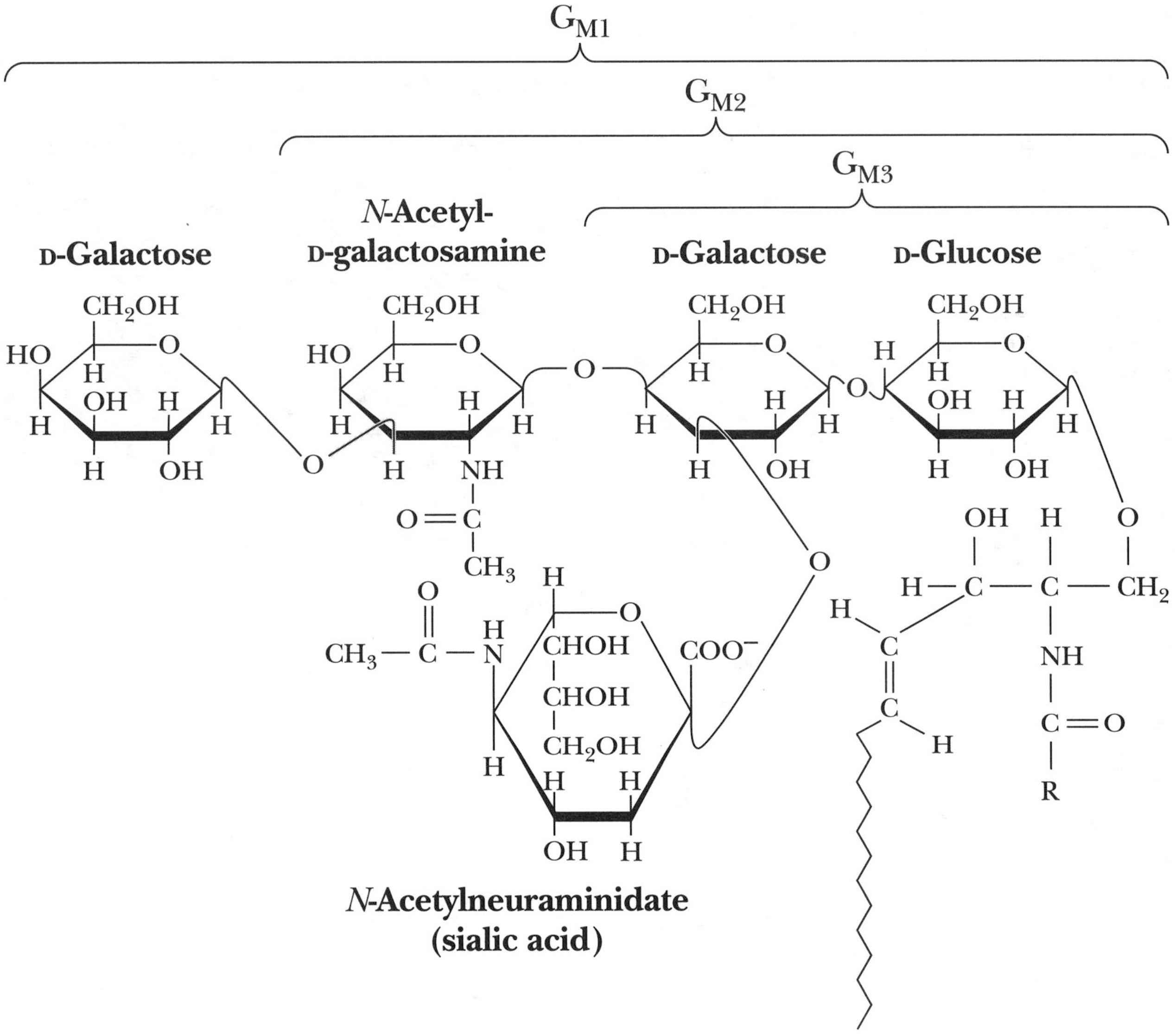

Gangliosides G_{M1}, G_{M2}, and G_{M3}

Gangliosides

Figure 7.8 Ganglioside structures

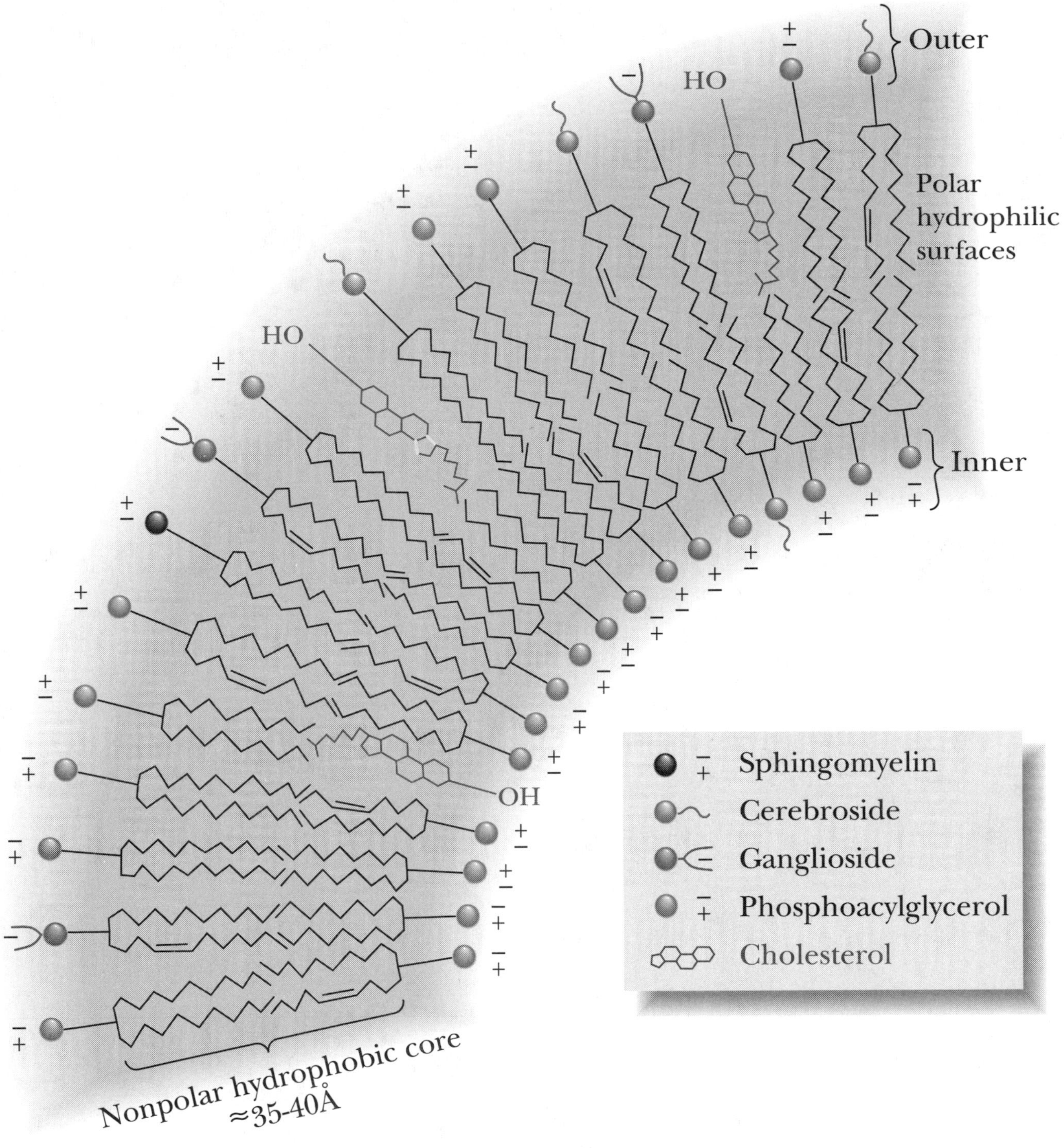

Outer
HO
Polar
hydrophilic
surfaces
HO
Inner
OH
Nonpolar hydrophobic core
≈35-40Å
Sphingomyelin
Cerebroside
Ganglioside
Phosphoacylglycerol
Cholesterol

Figure 7.11 Lipid bilayer asymmetry

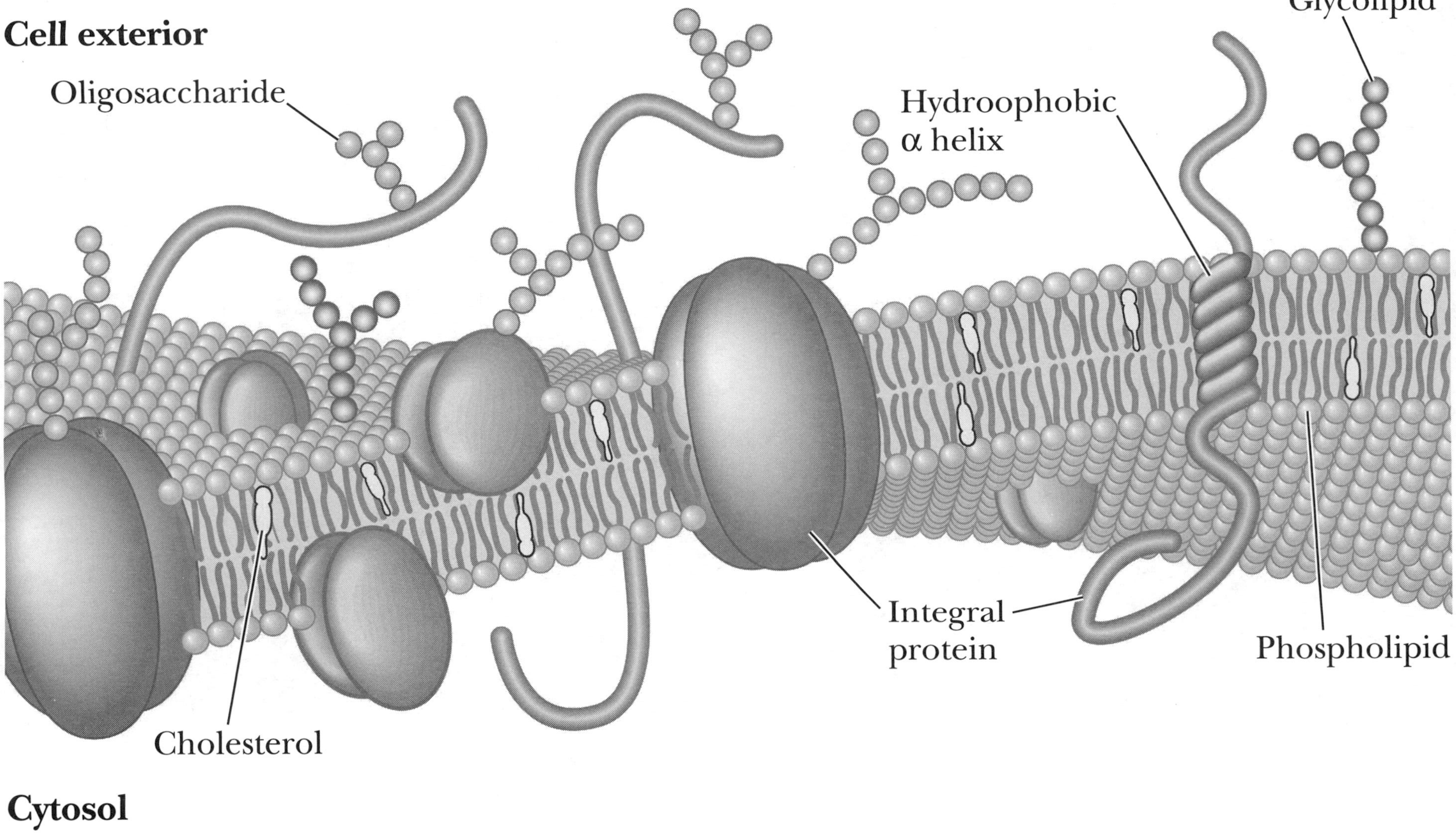

Figure 7.18 Fluid mosaic model for membrane structure

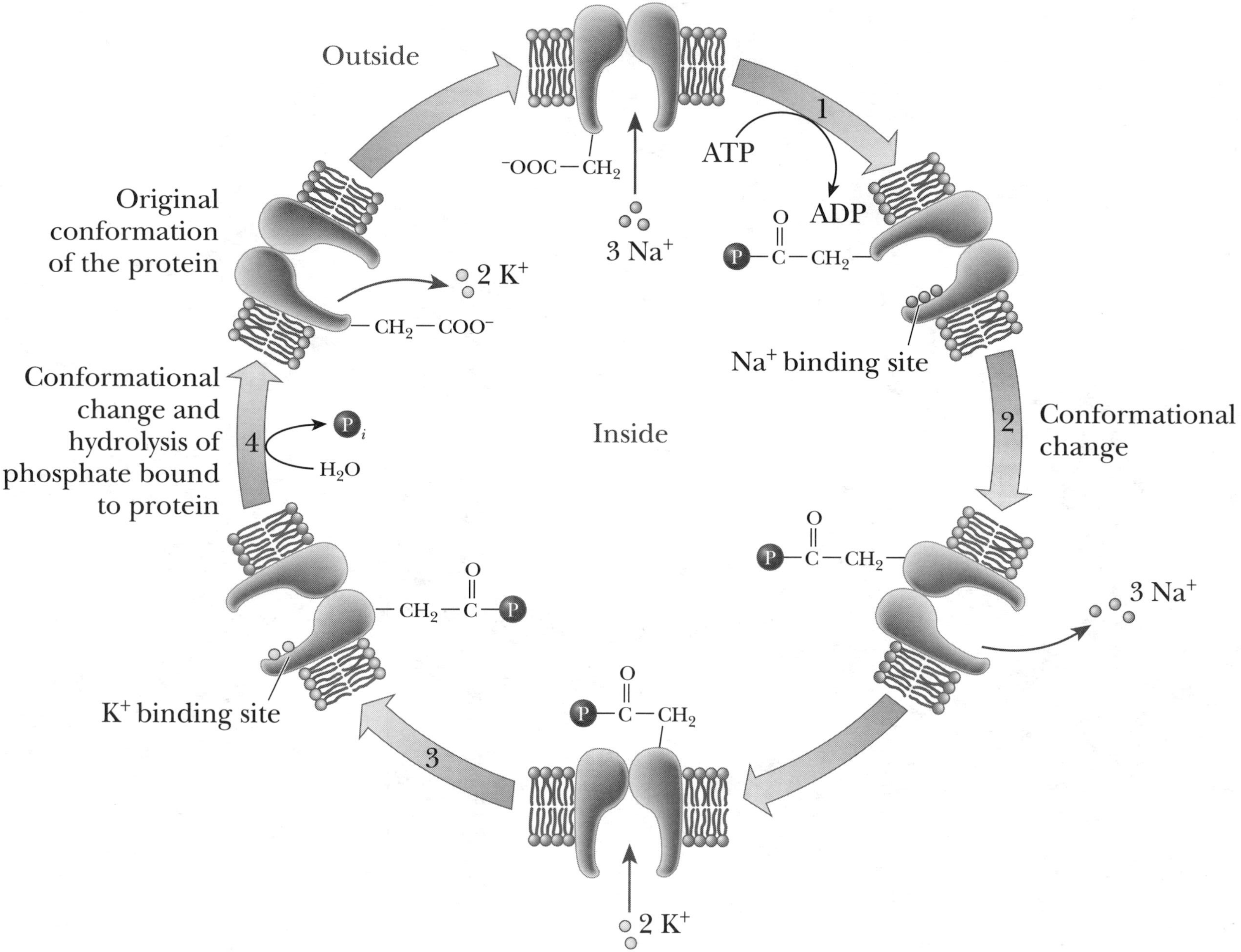

Outside
$^{-}OOC-CH_2$
3 Na^+
1
ATP
ADP
P—C(=O)—CH_2—
Na^+ binding site
2
Conformational change
P—C(=O)—CH_2—
3 Na^+
P—C(=O)—CH_2
2 K^+
3
K^+ binding site
—CH_2—C(=O)—P
Conformational change and hydrolysis of phosphate bound to protein
4
P_i
H_2O
Original conformation of the protein
2 K^+
—CH_2—COO^-
Inside

Figure 7.24 The sodium-potassium ion pump

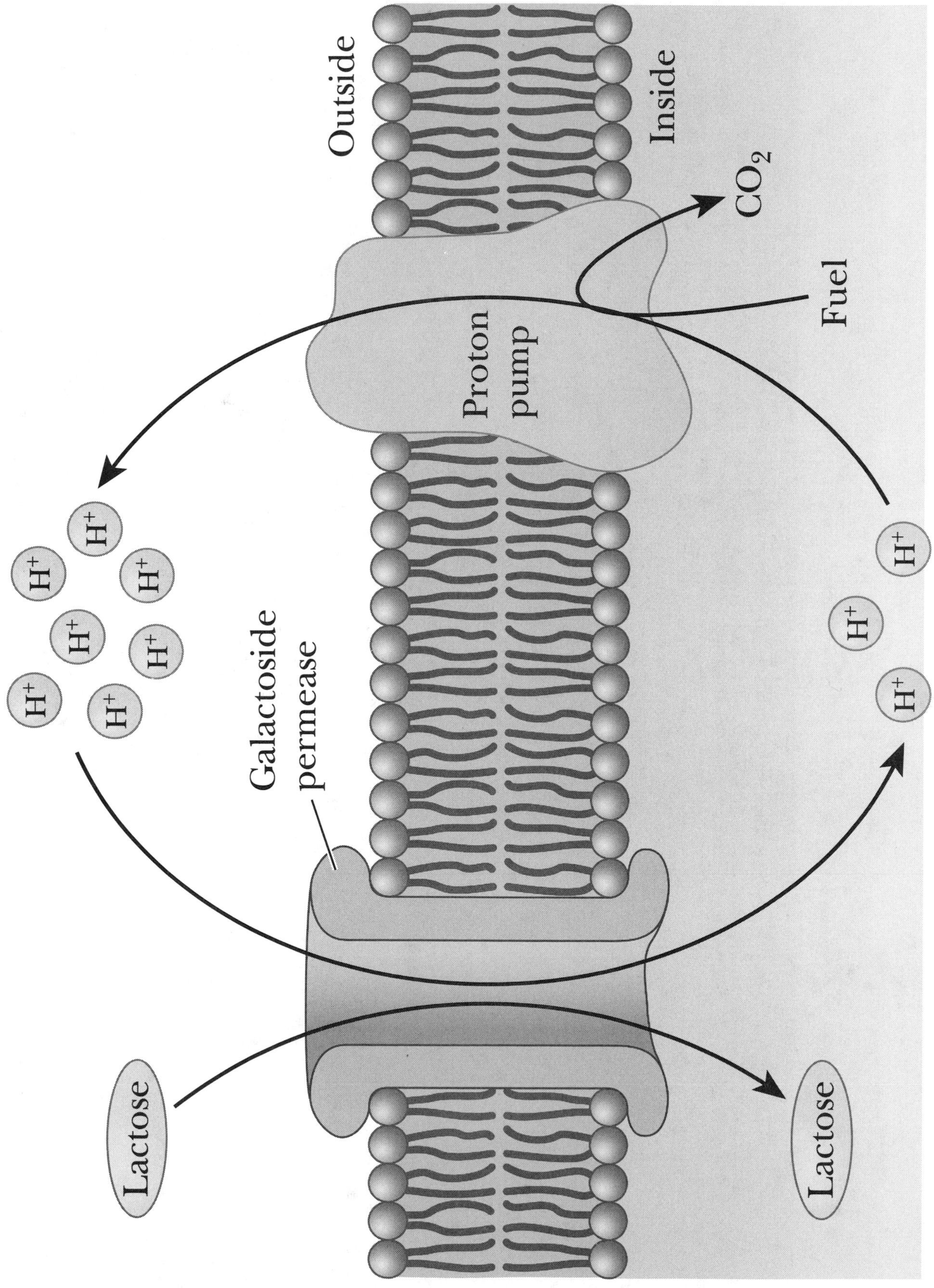

H^+
H^+
H^+
H^+
H^+
H^+
H^+
Lactose
Galactoside permease
Outside
Proton pump
Inside
CO_2
Lactose
Fuel
H^+
H^+
H^+

Figure 7.26 Secondary active transport

(a)

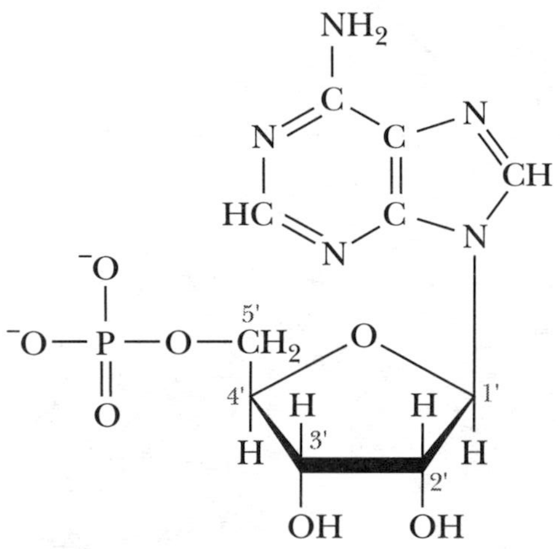

Adenosine 5'-monophosphate

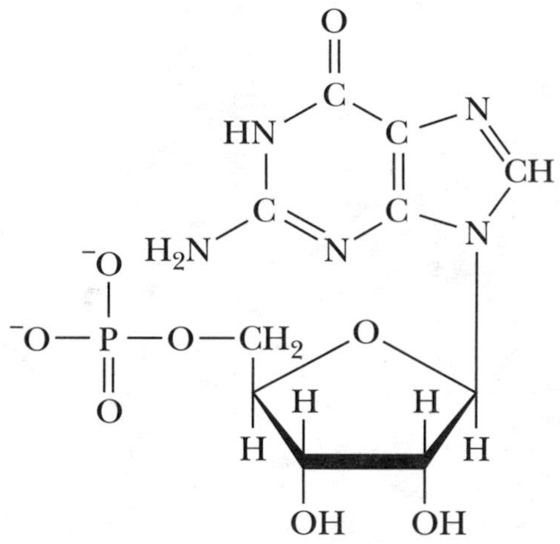

Guanosine 5'-monophosphate

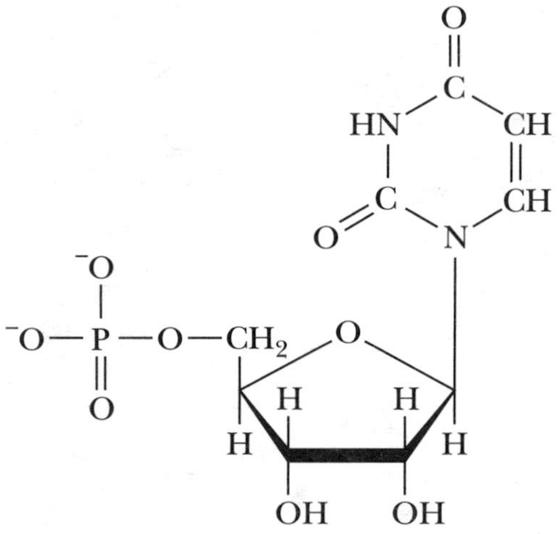

Uridine 5'-monophosphate

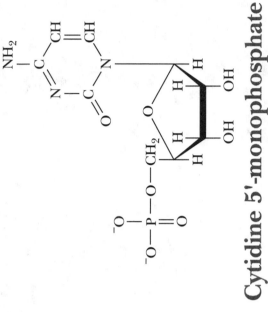

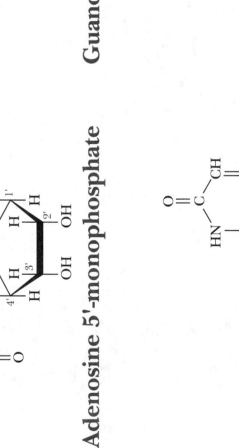

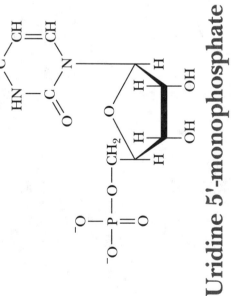

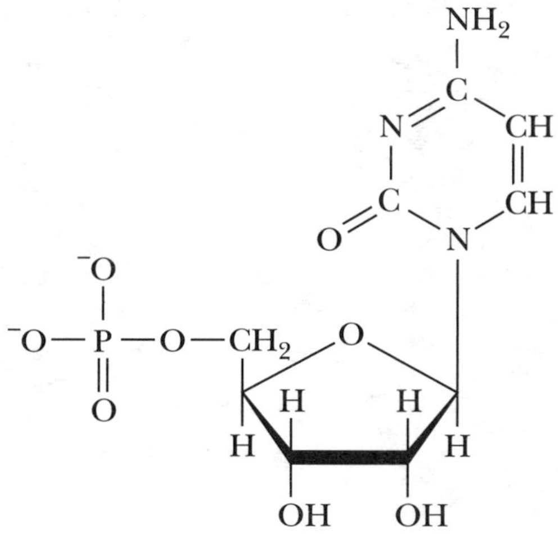

Cytidine 5'-monophosphate

Figure 8.4a Commonly occurring nucleotides

(b)

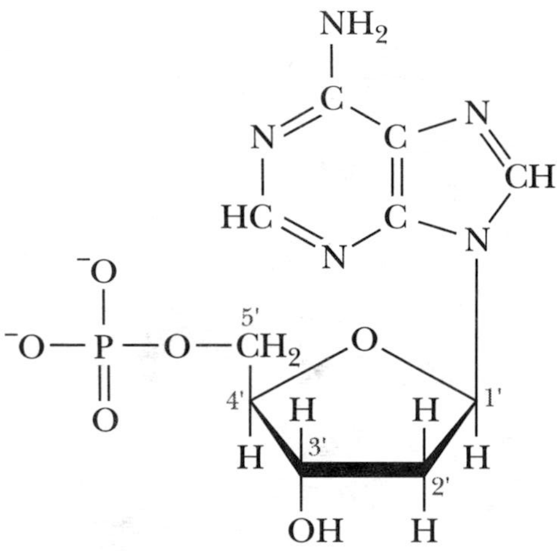

Deoxyadenosine 5'-monophosphate

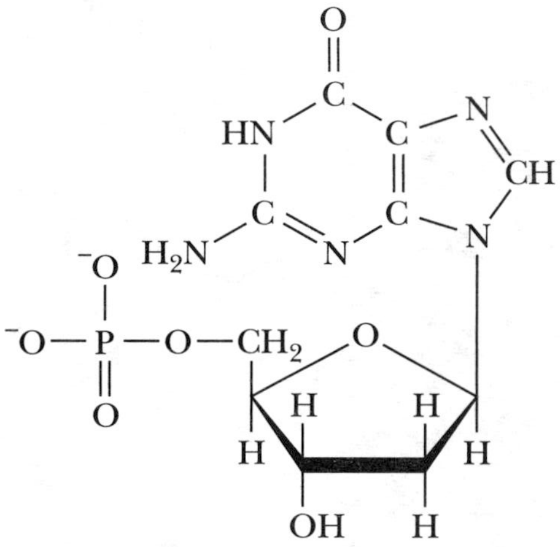

Deoxyguanosine 5'-monophosphate

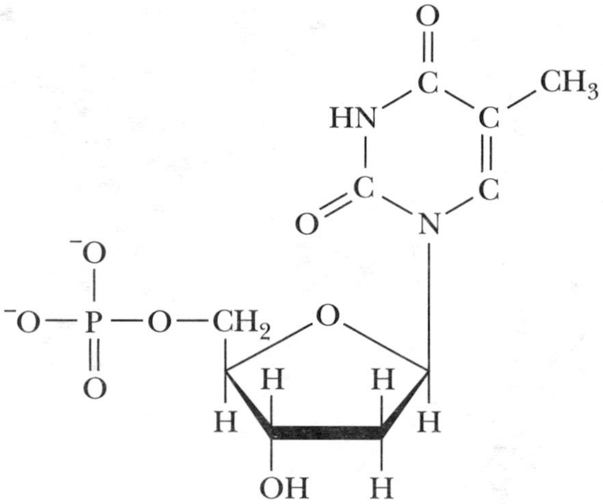

Deoxythymidine 5'-monophosphate

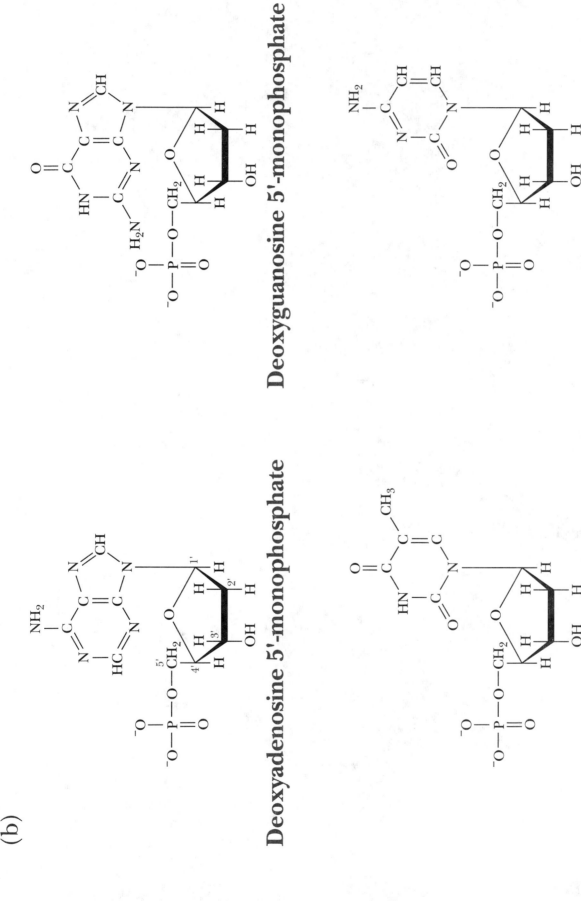

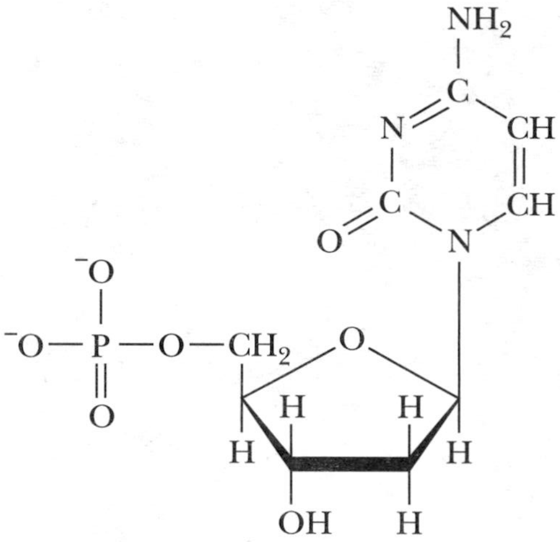

Deoxycytidine 5'-monophosphate

Figure 8.4b Commonly occurring nucleotides

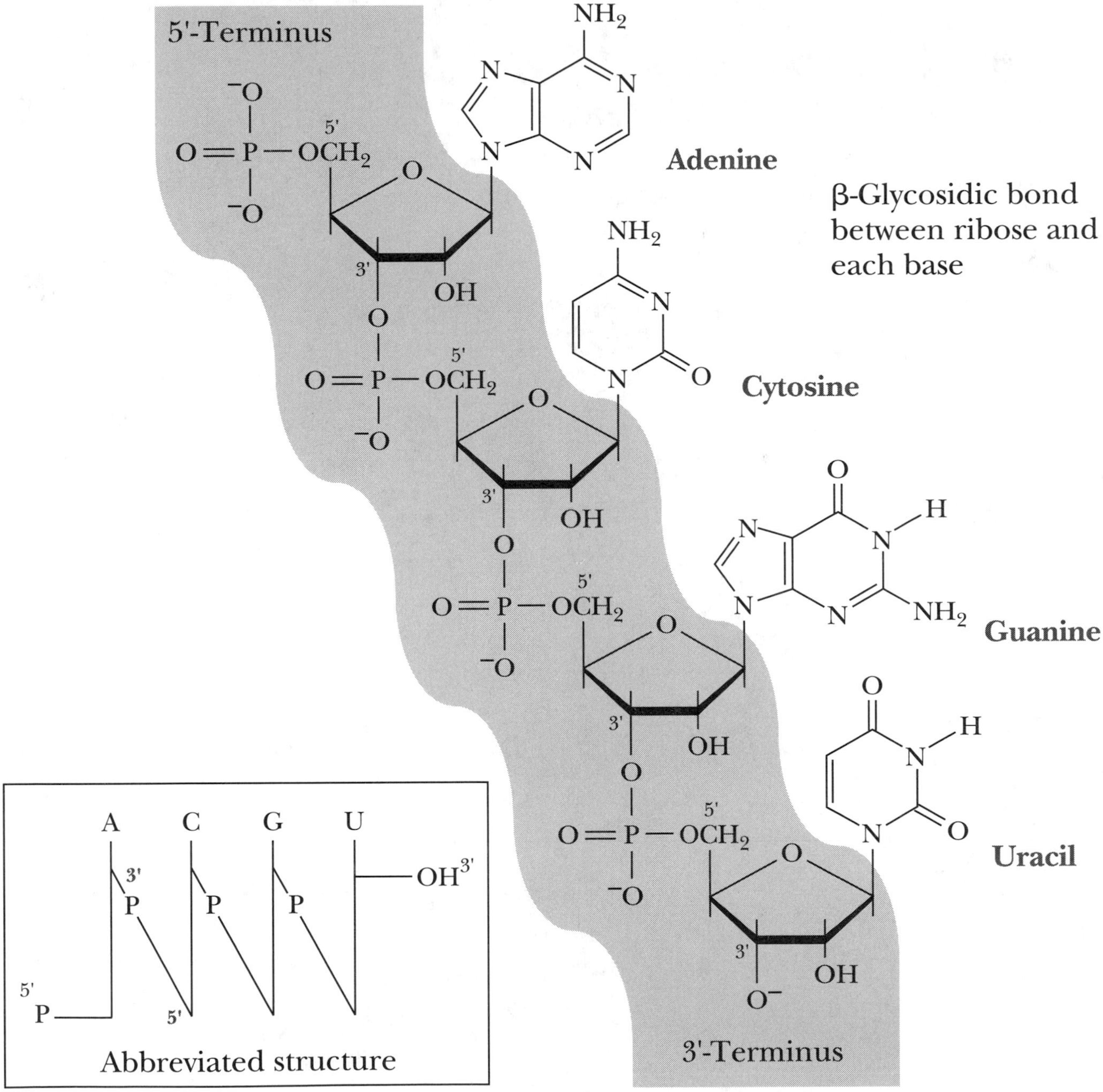
5'-Terminus
Adenine
β-Glycosidic bond between ribose and each base
Cytosine
Guanine
Uracil
3'-Terminus
A
C
G
U
Abbreviated structure

Figure 8.5 A fragment of an RNA chain

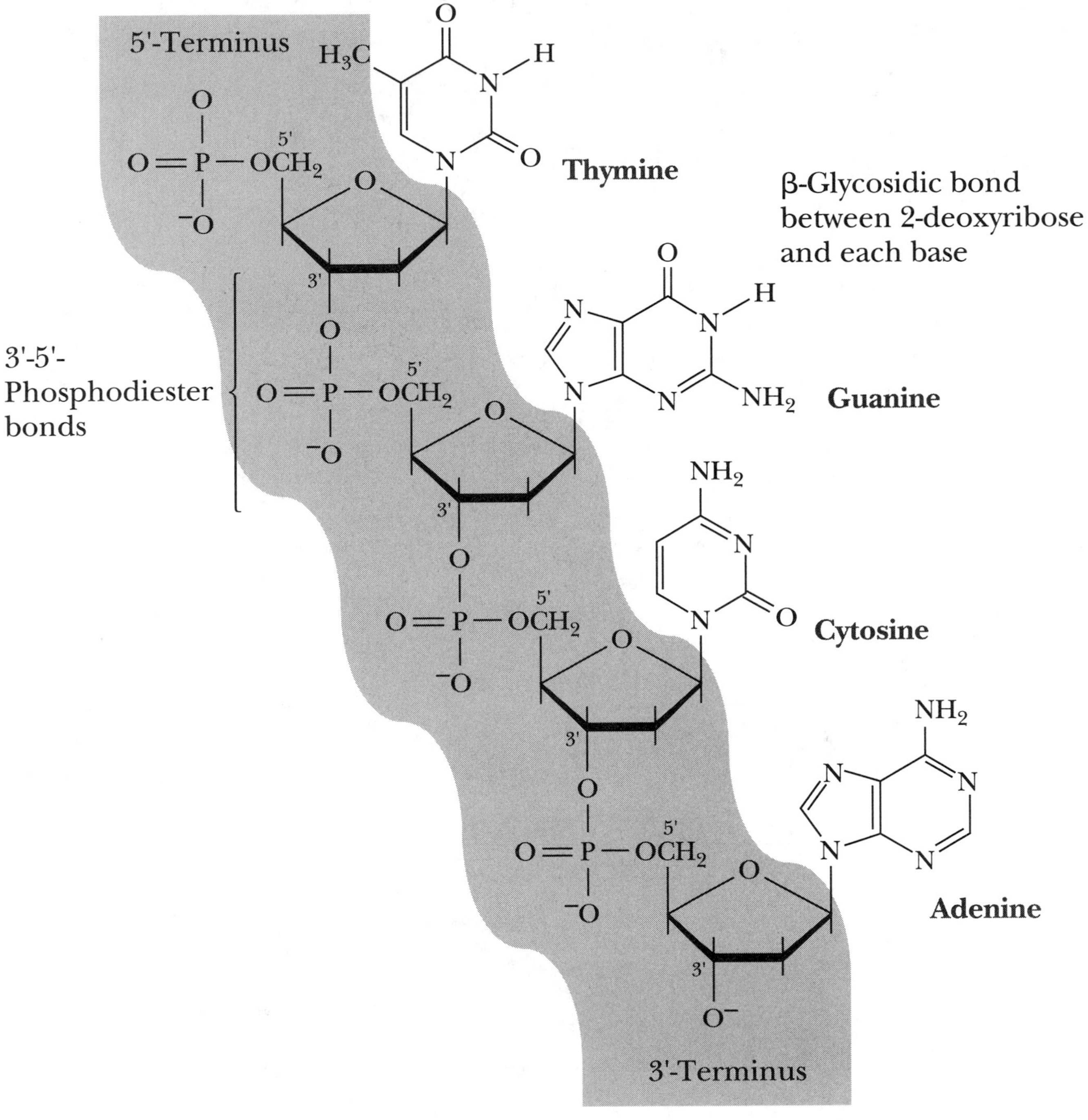
5'-Terminus
Thymine
β-Glycosidic bond between 2-deoxyribose and each base
Guanine
3'-5'-Phosphodiester bonds
Cytosine
Adenine
3'-Terminus

Figure 8.6 A portion of a DNA chain

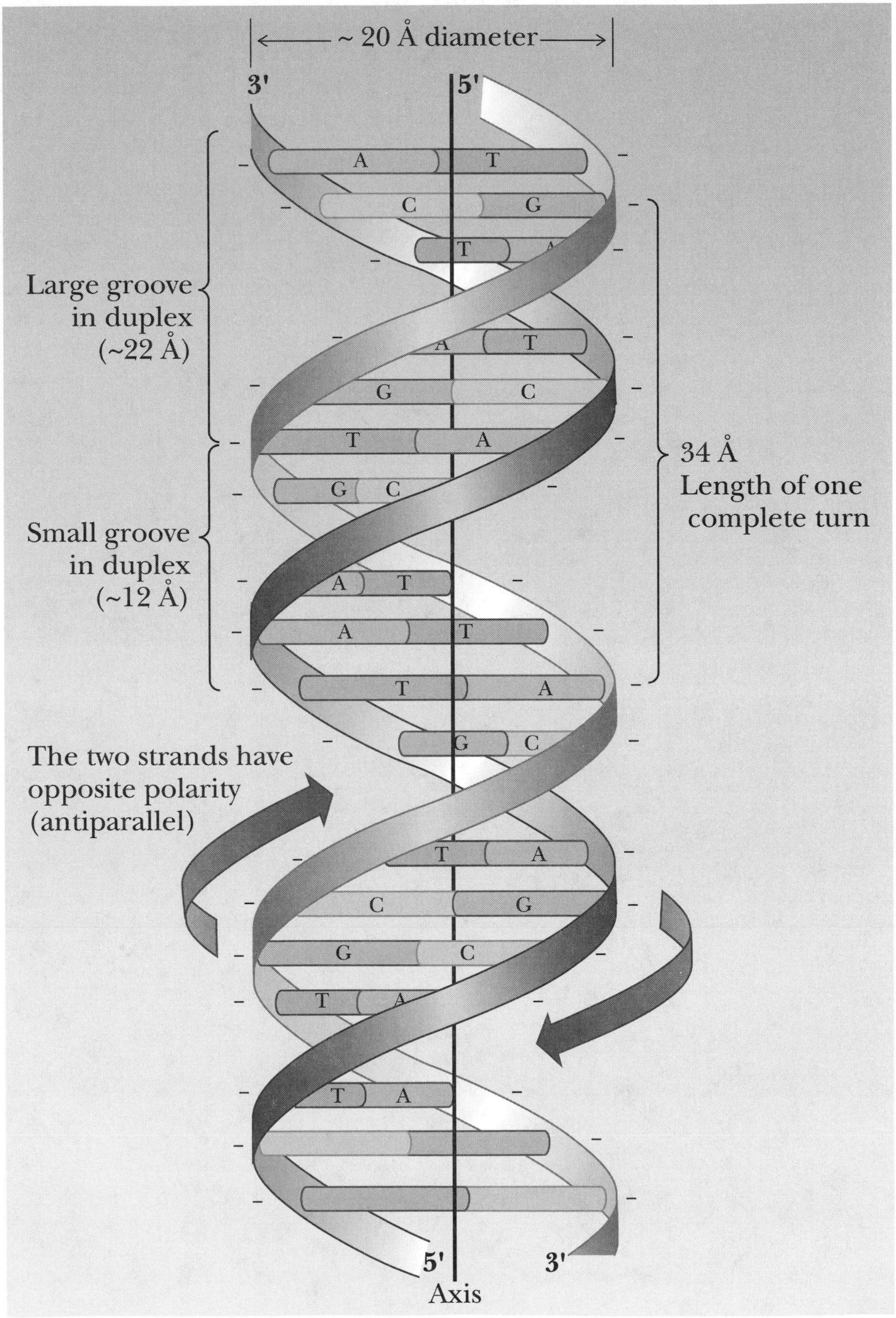
~ 20 Å diameter
3'
5'
Large groove in duplex (~22 Å)
Small groove in duplex (~12 Å)
34 Å Length of one complete turn
The two strands have opposite polarity (antiparallel)
5'
3'
Axis

Figure 8.7 The double helix

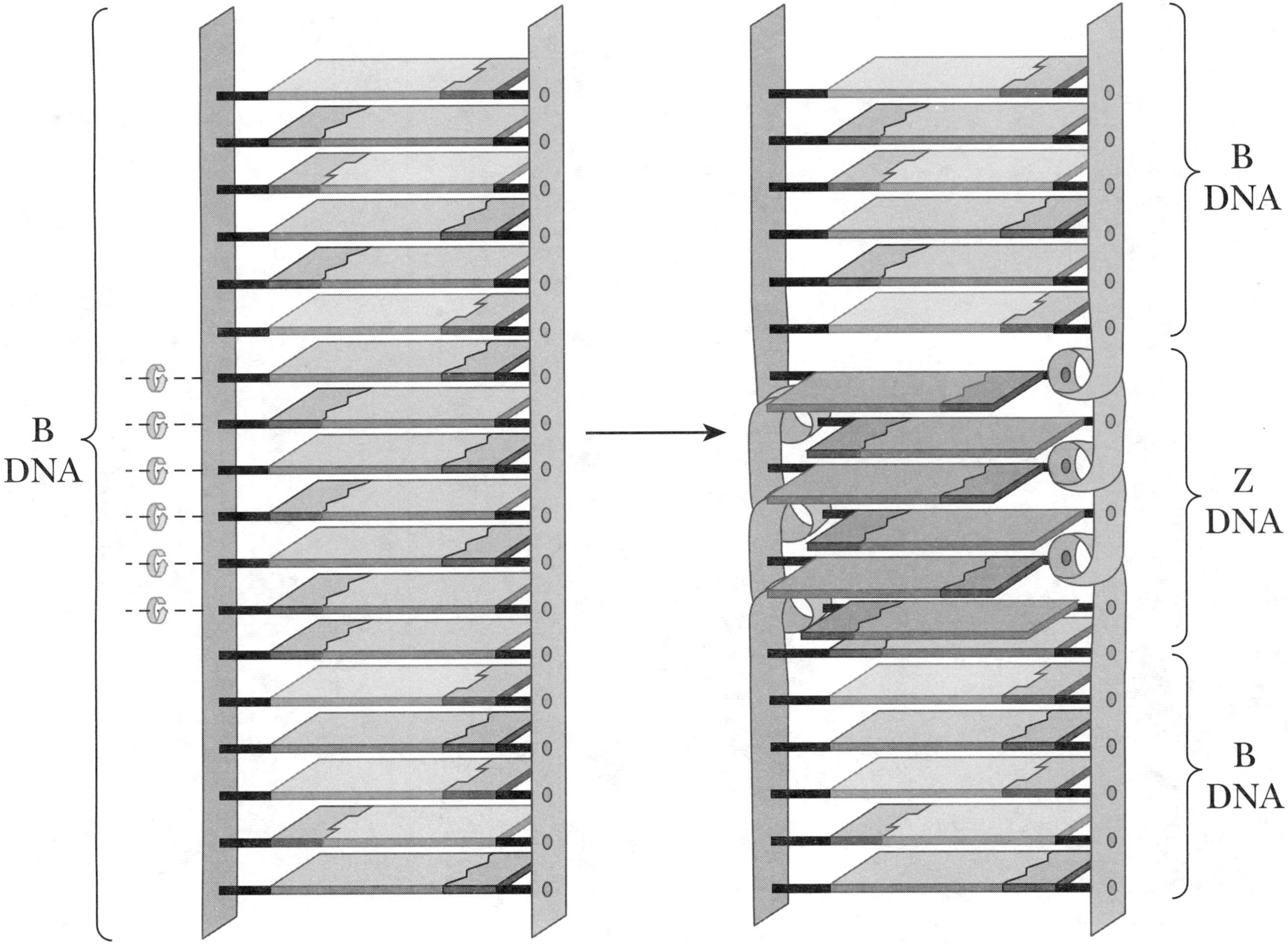

Figure 8.11 A Z-DNA section in the middle of a B-DNA

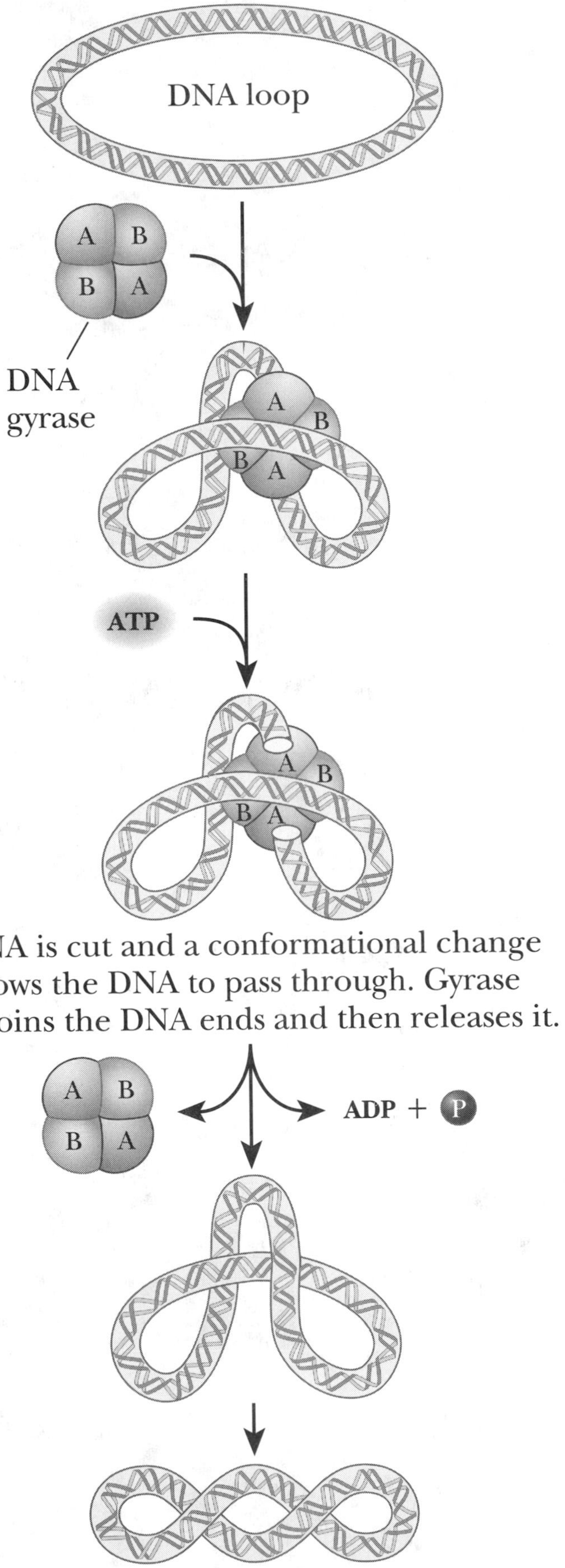

DNA loop
A
B
B
A
DNA gyrase
ATP
DNA is cut and a conformational change allows the DNA to pass through. Gyrase rejoins the DNA ends and then releases it.
ADP + P

Figure 8.15 A model for the action of bacterial DNA gyrase

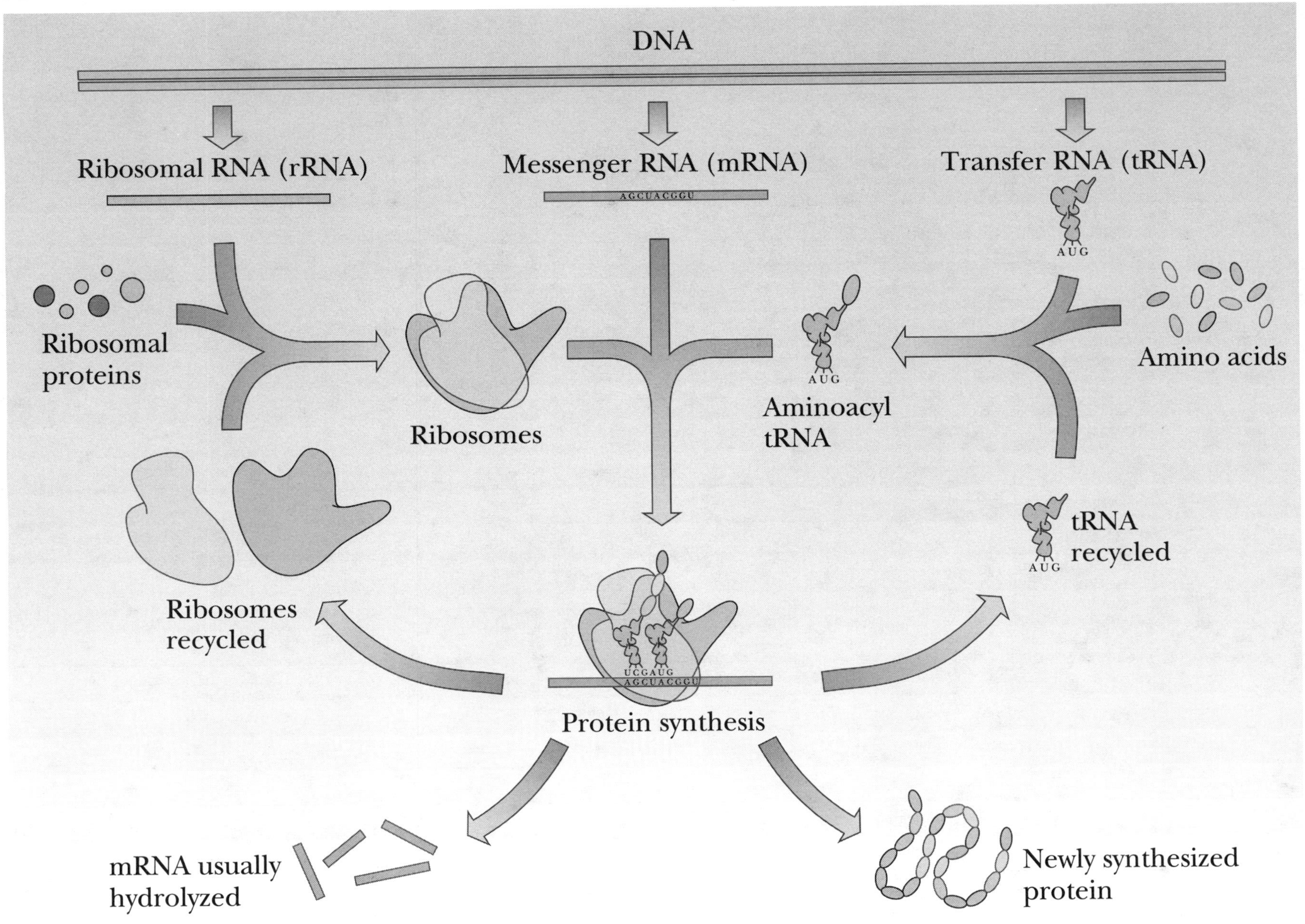

DNA
Ribosomal RNA (rRNA)
Messenger RNA (mRNA)
AGCUACGGU
Transfer RNA (tRNA)
AUG
Ribosomal proteins
Ribosomes
AUG
Aminoacyl tRNA
Amino acids
tRNA recycled
AUG
Ribosomes recycled
UCGAUG
AGCUACGGU
Protein synthesis
mRNA usually hydrolyzed
Newly synthesized protein

Figure 8.19 Types of RNA

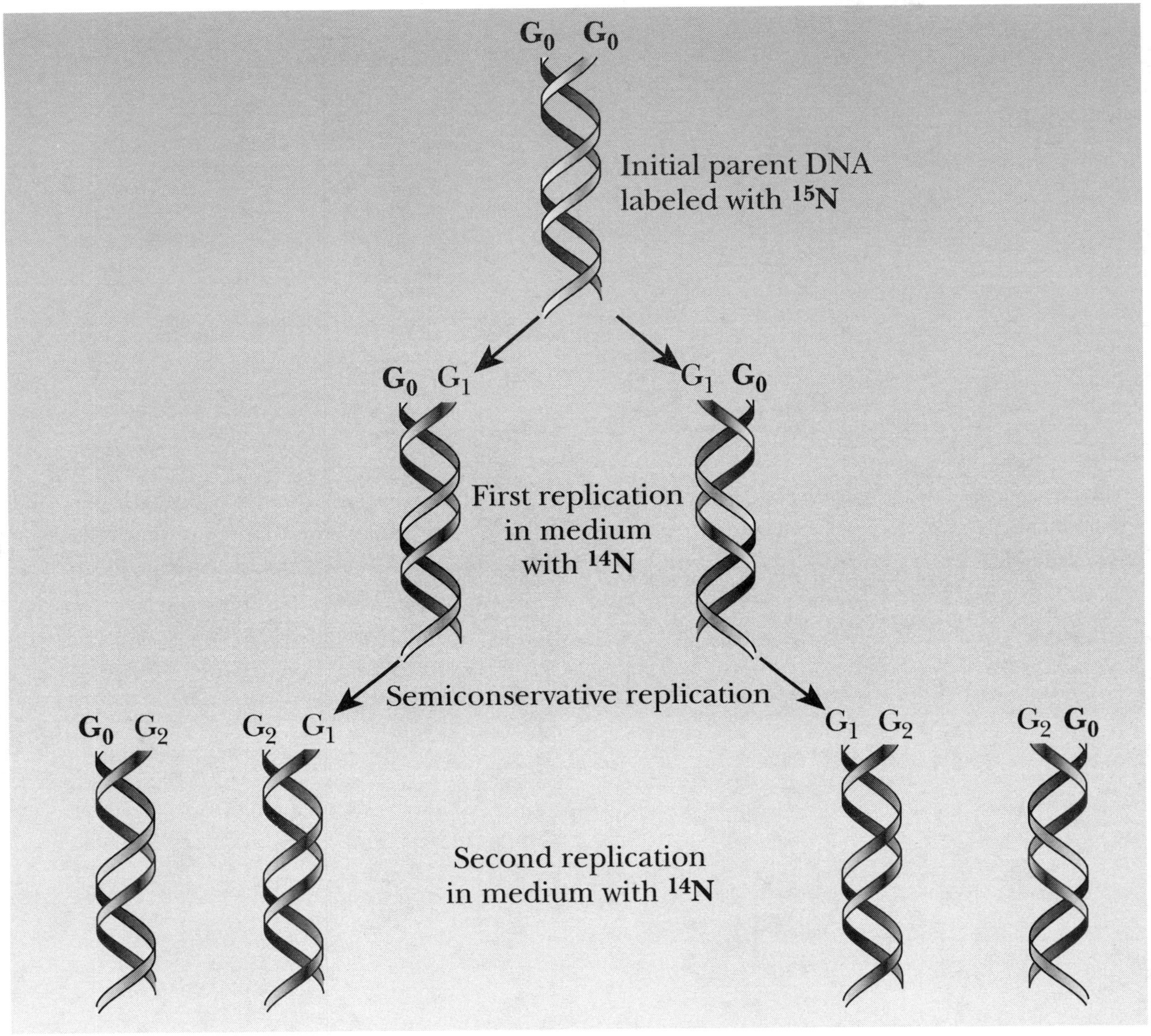

G_0 G_0
Initial parent DNA labeled with ^{15}N
G_0 G_1
G_1 G_0
First replication in medium with ^{14}N
Semiconservative replication
G_0 G_2
G_2 G_1
G_1 G_2
G_2 G_0
Second replication in medium with ^{14}N

Figure 9.2 Semiconservative replication

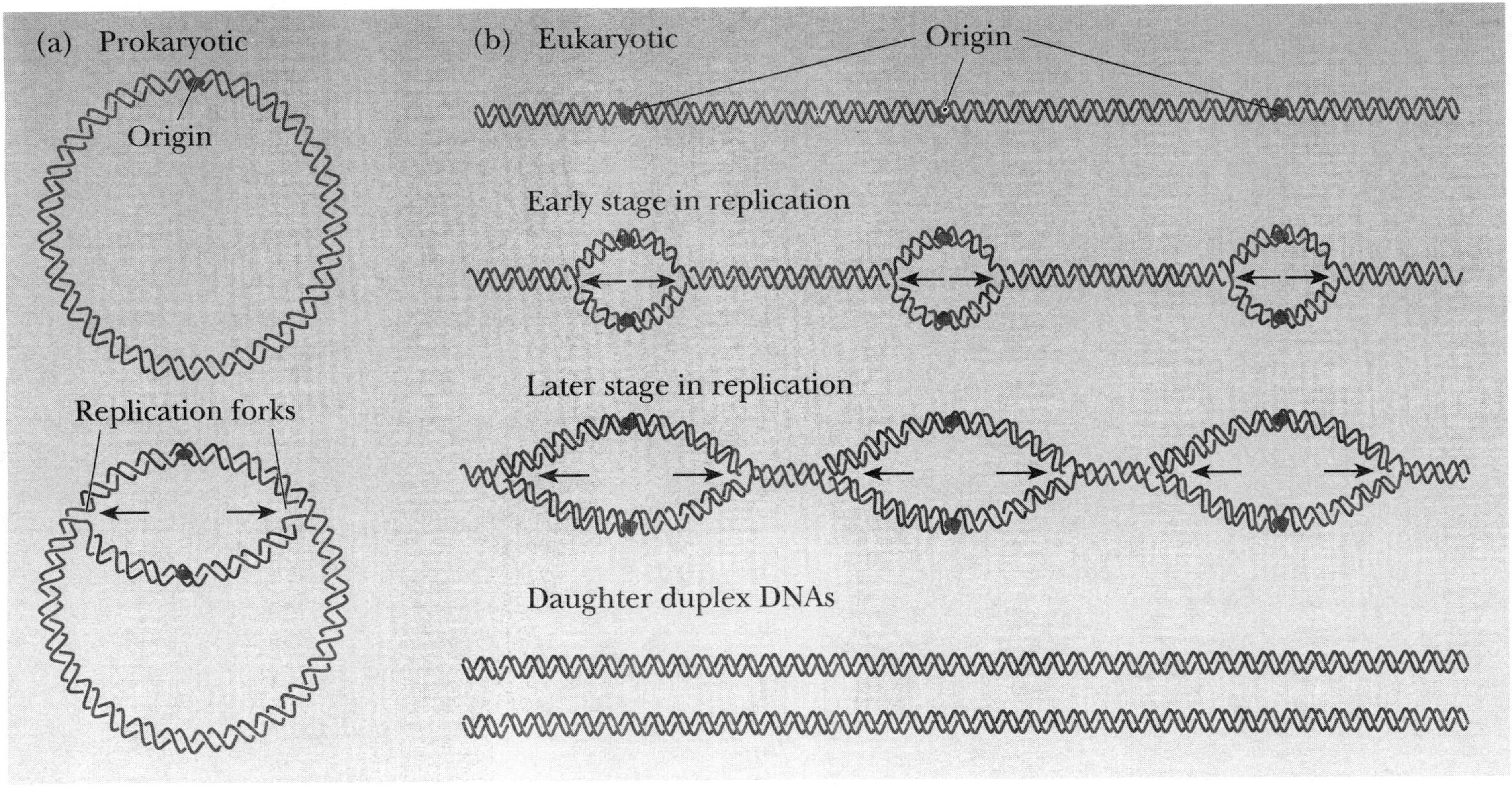

Figure 9.4 Bidirectional replication of DNA in prokaryotes and eukaryotes

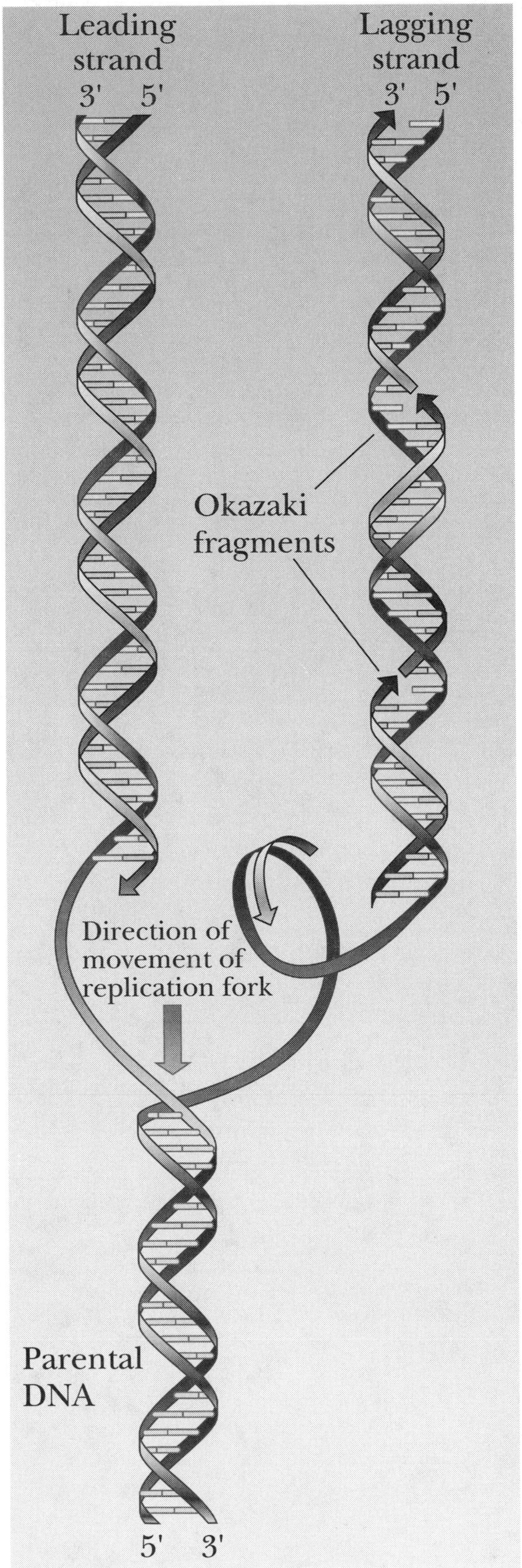

Figure 9.5 The semidiscontinuous model for DNA replication

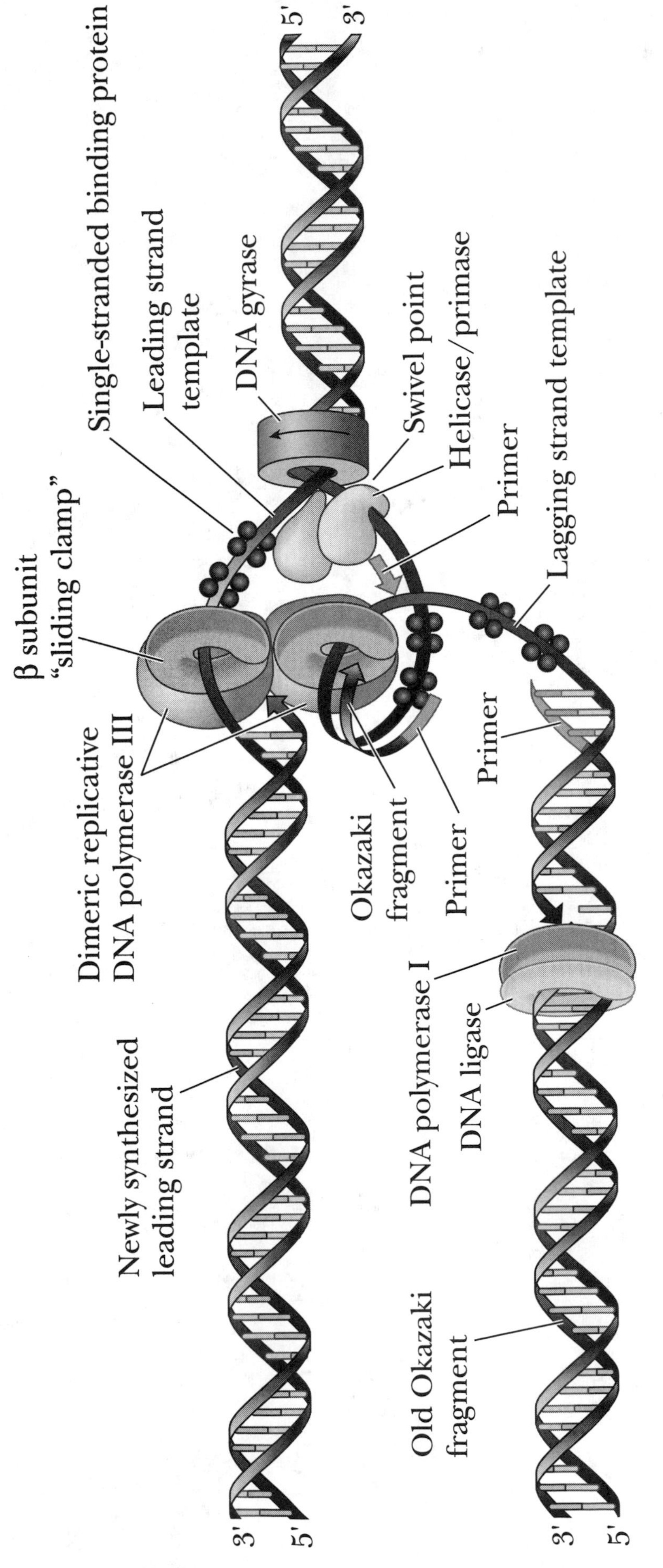

Figure 9.10 General features of the replication fork

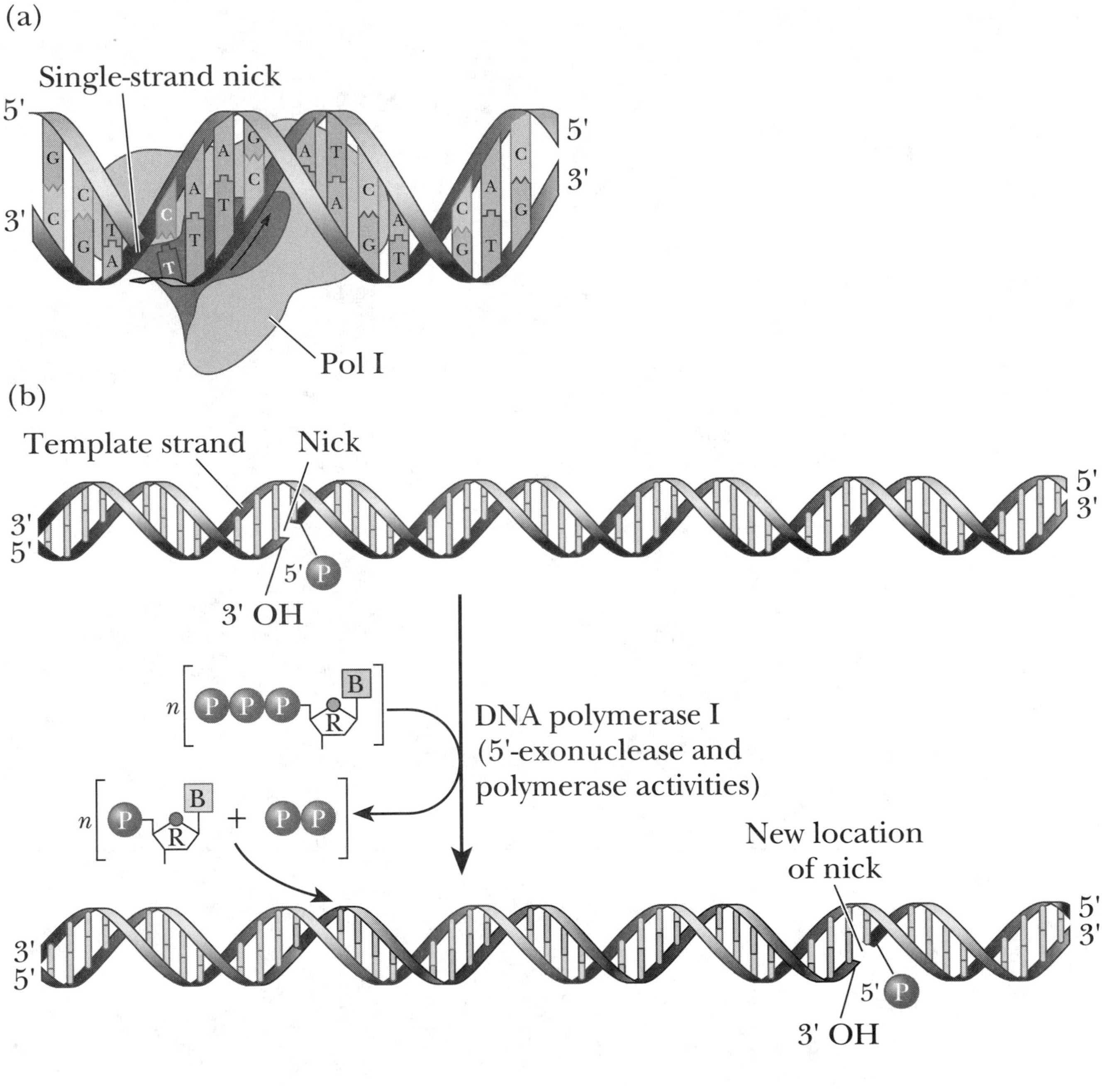

Figure 9.12 The 5′ → 3′ exonuclease activity of DNA polymerase I

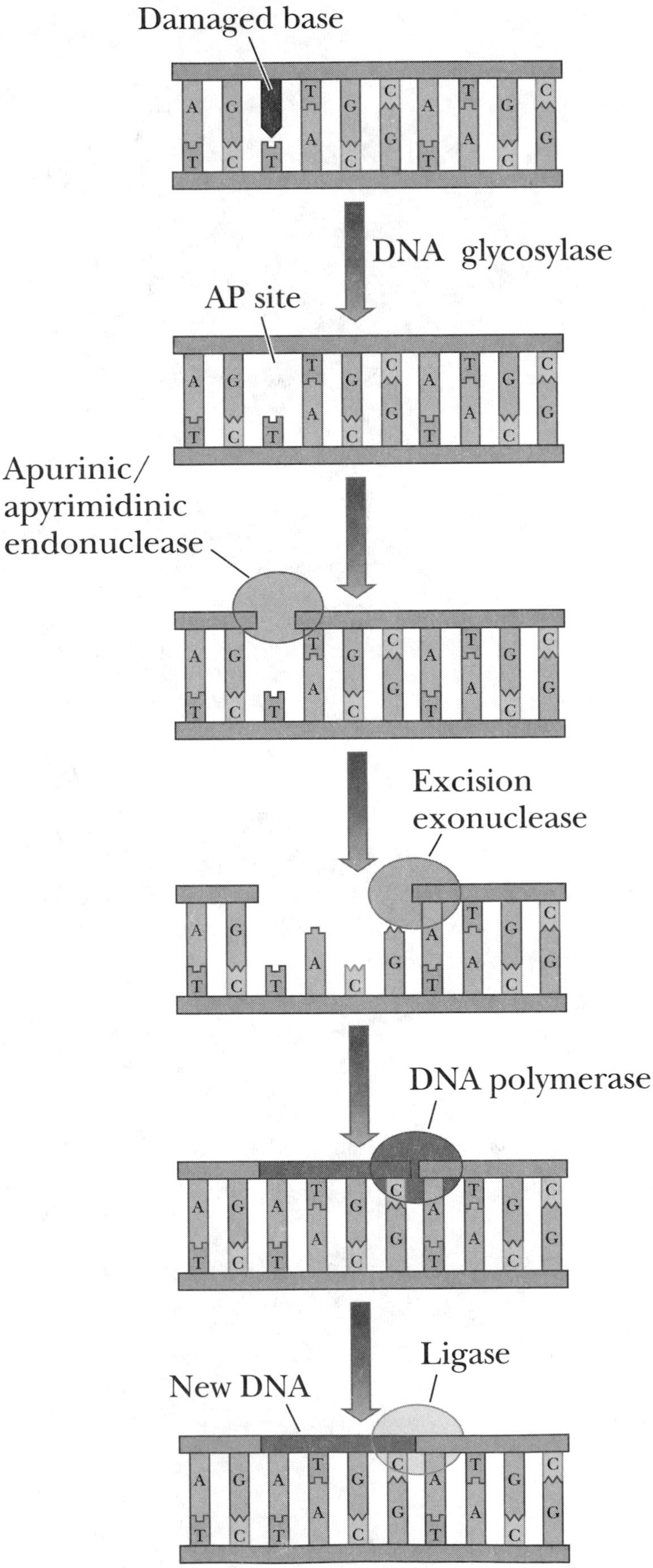

Damaged base
DNA glycosylase
AP site
Apurinic/
apyrimidinic
endonuclease
Excision
exonuclease
DNA polymerase
New DNA
Ligase

Figure 9.16 Base excision repair

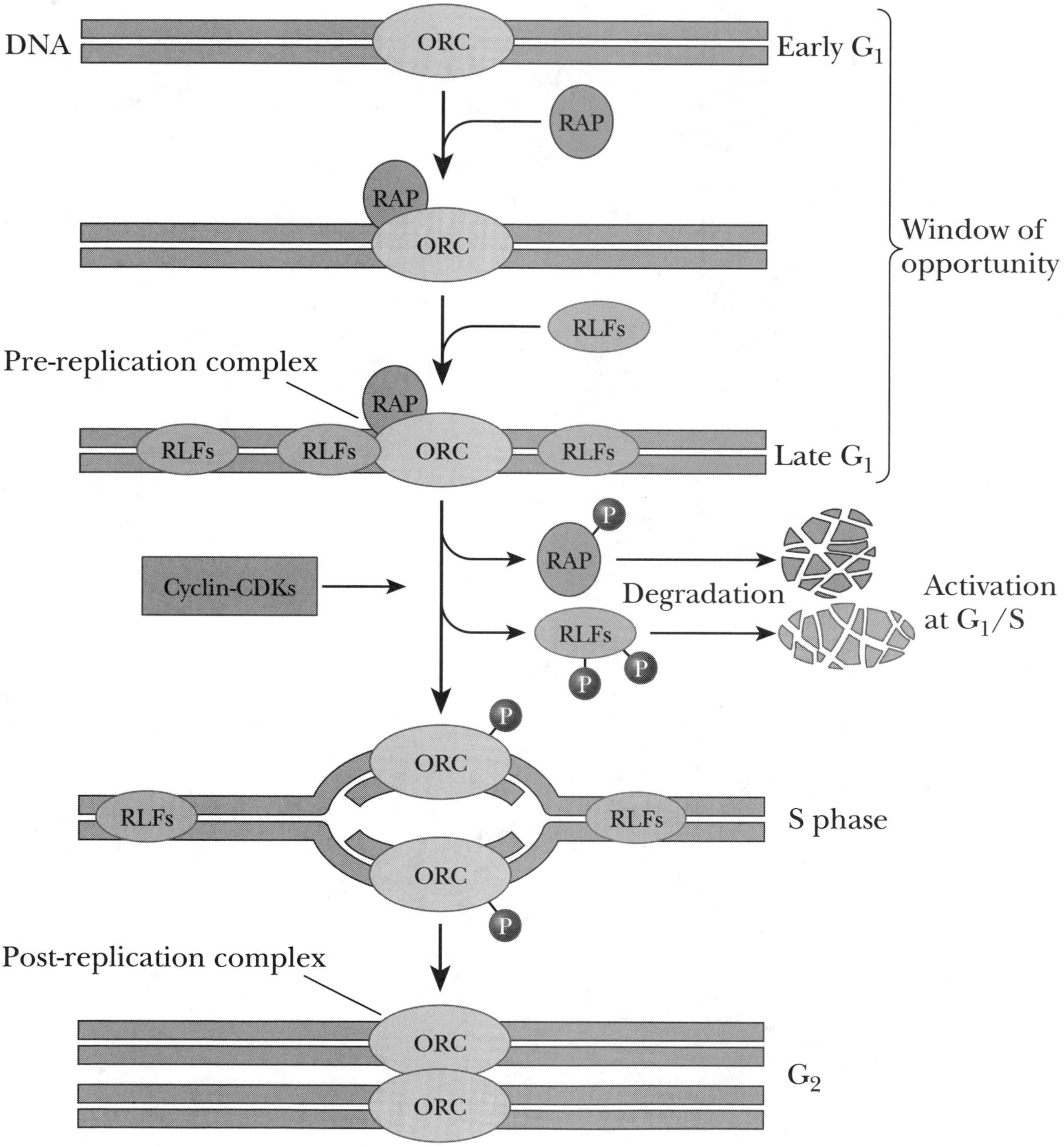

Figure 9.19 Model for the initiation of the DNA replication cycle in eukaryotes

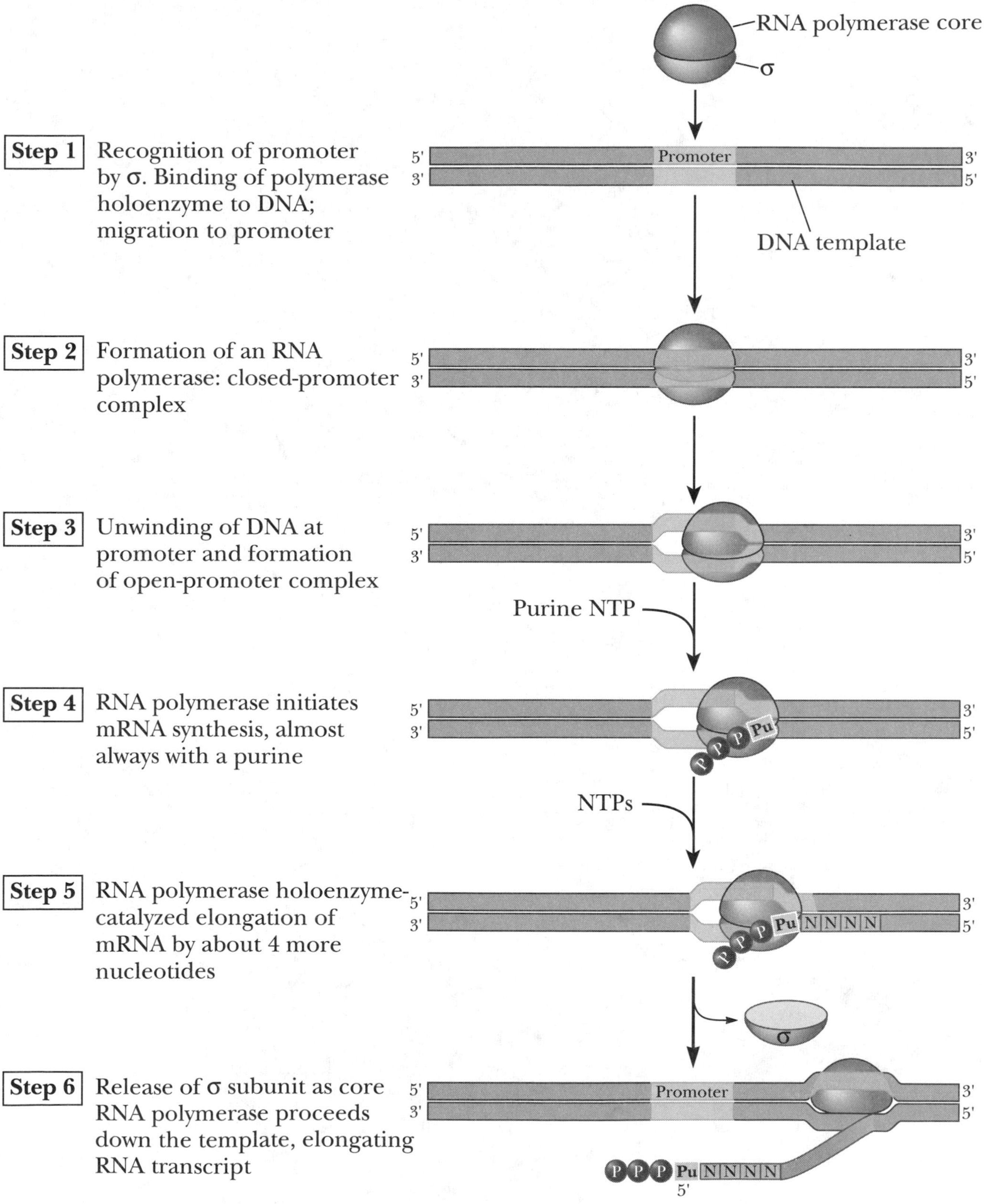
RNA polymerase core
σ
Step 1
Recognition of promoter by σ. Binding of polymerase holoenzyme to DNA; migration to promoter
5'
3'
Promoter
3'
5'
DNA template
Step 2
Formation of an RNA polymerase: closed-promoter complex
Step 3
Unwinding of DNA at promoter and formation of open-promoter complex
Purine NTP
Step 4
RNA polymerase initiates mRNA synthesis, almost always with a purine
P P P Pu
NTPs
Step 5
RNA polymerase holoenzyme-catalyzed elongation of mRNA by about 4 more nucleotides
P P P Pu N N N N
σ
Step 6
Release of σ subunit as core RNA polymerase proceeds down the template, elongating RNA transcript
Promoter
P P P Pu N N N N
5'

Figure 10.3 The basic order of events in prokaryotic transcription initiation and elongation

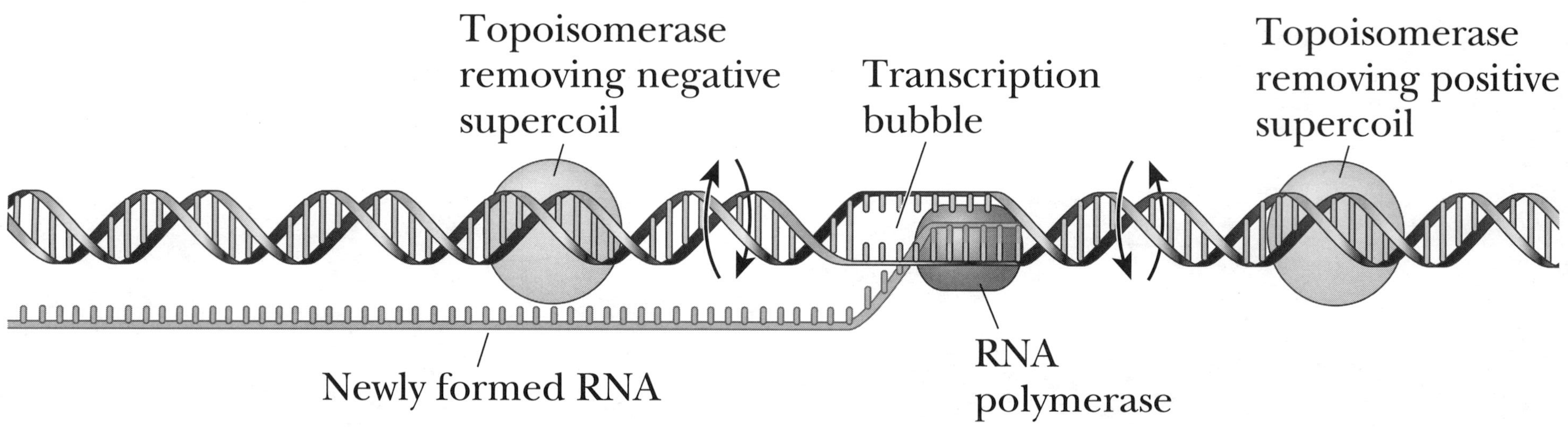

Figure 10.4 Topoisomerases remove supercoils that would form ahead of and behind the transcription bubble

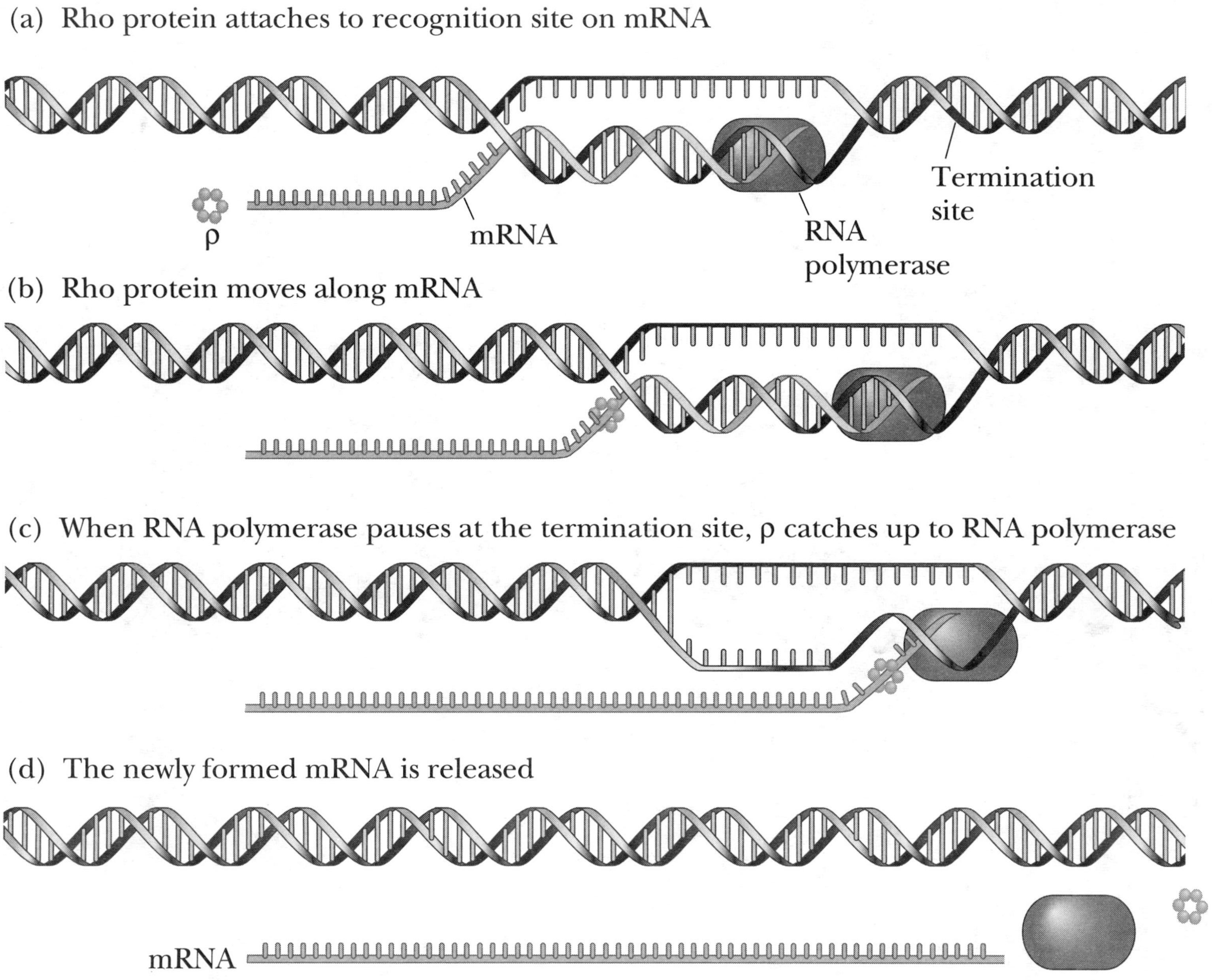

(a) Rho protein attaches to recognition site on mRNA
ρ
mRNA
RNA polymerase
Termination site
(b) Rho protein moves along mRNA
(c) When RNA polymerase pauses at the termination site, ρ catches up to RNA polymerase
(d) The newly formed mRNA is released
mRNA

Figure 10.6 Transcription termination by the rho factor

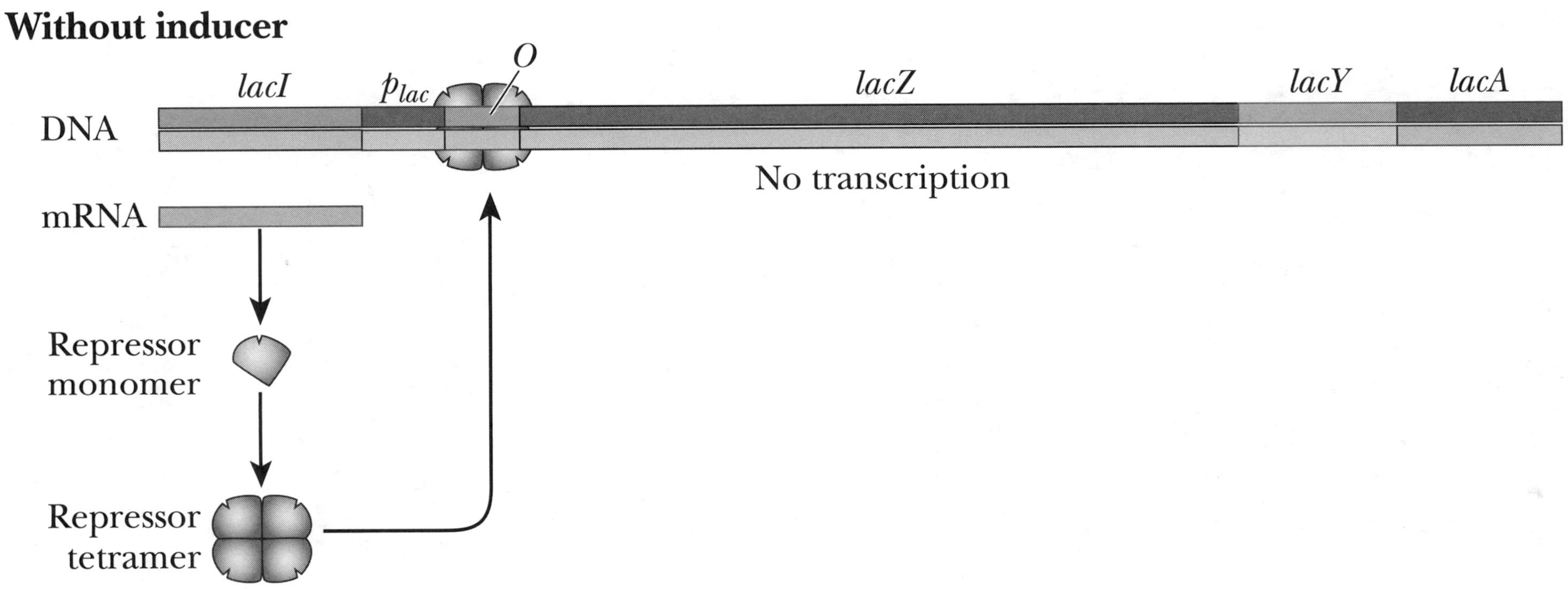

Figure 10.9 (top) The lacI gene produces a protein that represses the lac operon

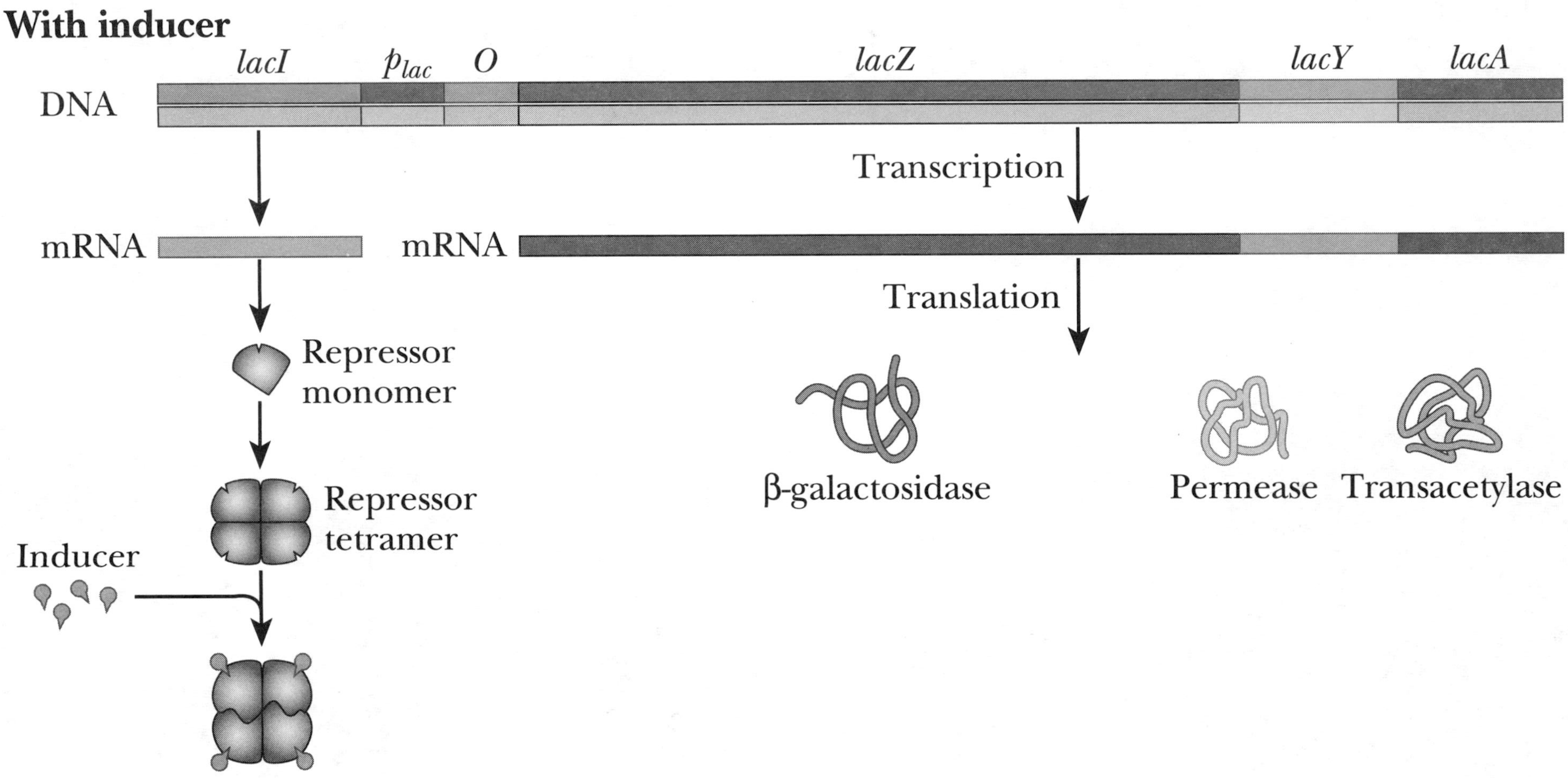

Figure 10.9 (bottom) The lacI gene produces a protein that represses the lac operon

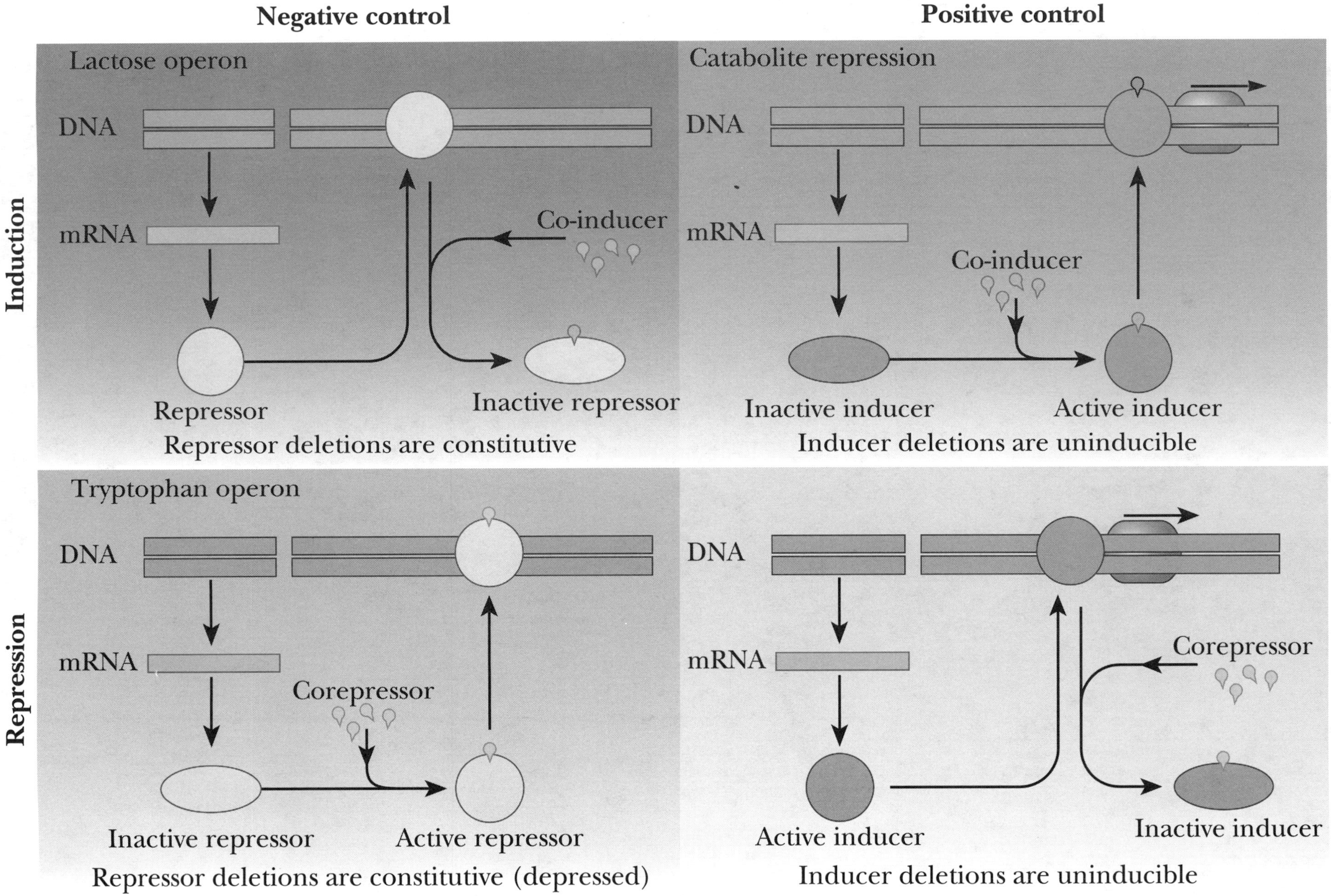

Figure 10.12 Basic control mechanisms seen in the expression of genes

trp p,O
trpL
trpE
trpD
trpC
trpB
trpA
DNA
Control sites
Attenuator
mRNA
Anthranilate synthase component I
Anthranilate synthase component II
Tryptophan synthase β subunit
Tryptophan synthase α subunit
1
Anthranilate synthase (CoI2CoII2)
2
N-(5'Phosphoribosyl)-anthranilate isomerase
Indole-3'-glycerol phosphate synthase
3
Tryptophan synthase (α2β2)
Chorismate
Glutamine
Glutamate + pyruvate
Anthranilate
PRPP
PPi
N-(5'-Phosphoribosyl)-anthranilate
Enol-1-o-carboxyphenylamino-1-deoxyribulose phosphate
CO2
Indole-3-glycerol-P
L-Serine
Glyceraldehyde-3-P
L-Tryptophan

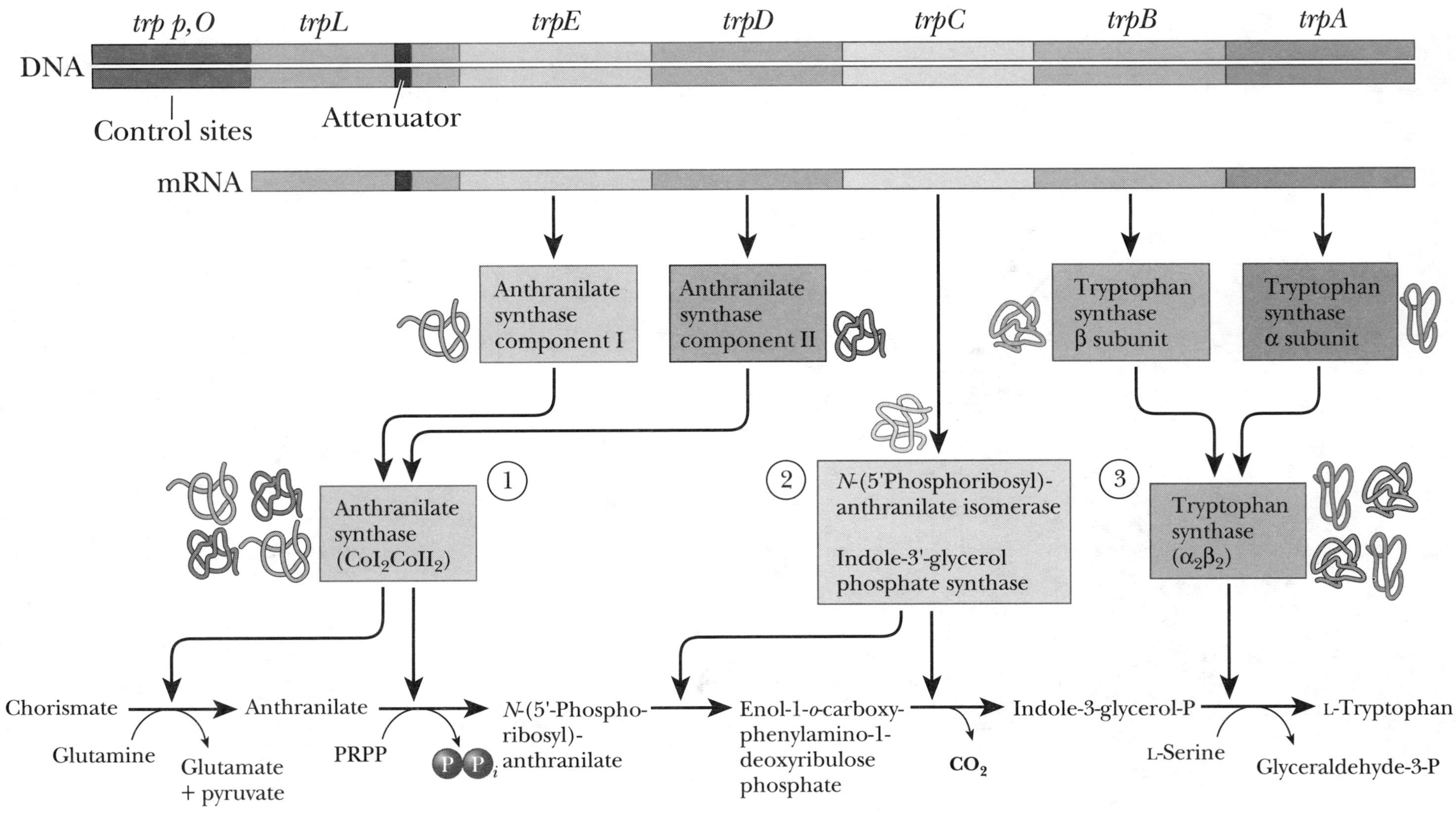

Figure 10.13 The trp operon of E. coli.

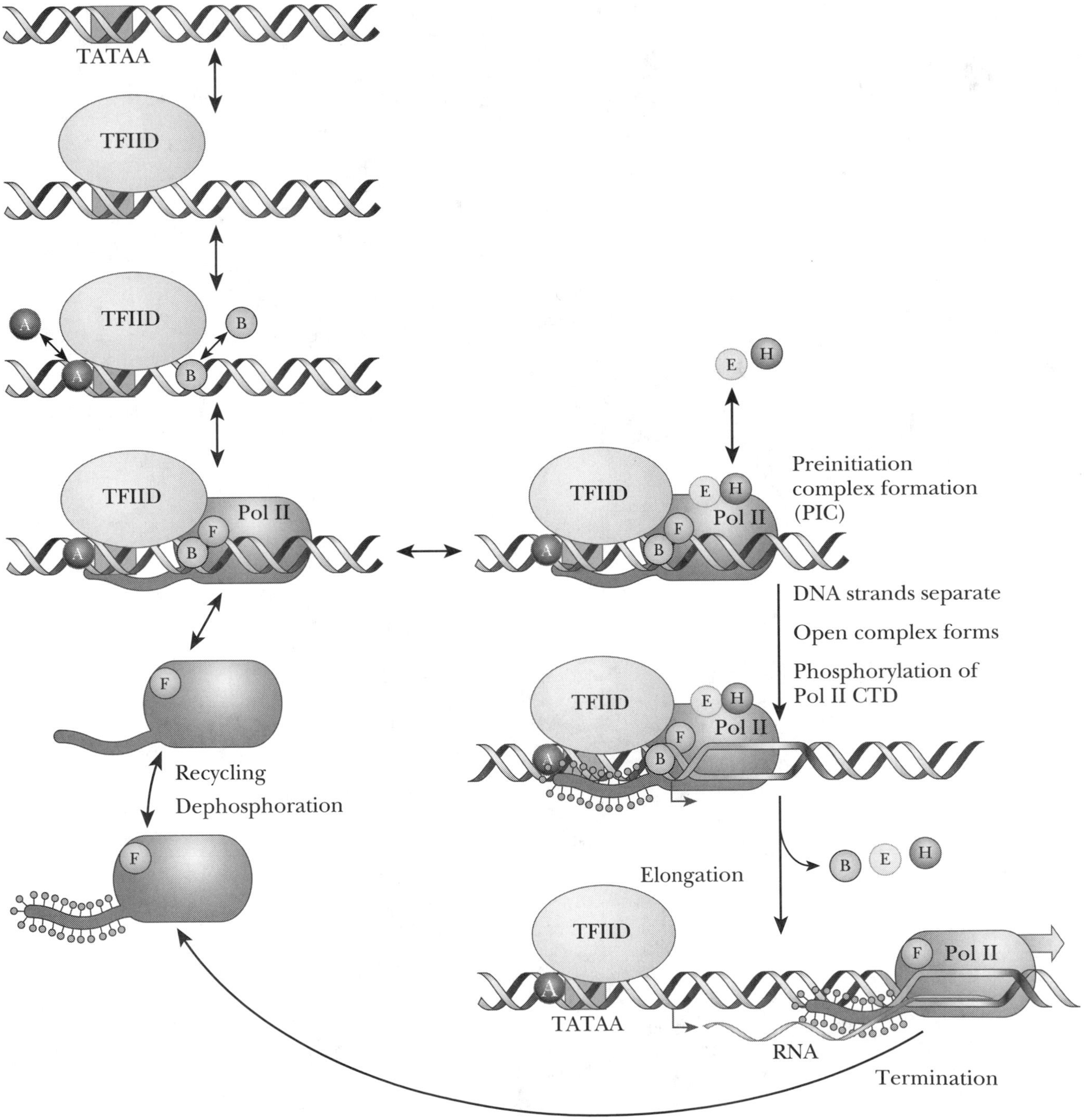

Figure 10.18 The order of events of transcription

Protein	Basic region A / Basic region B / Leucine zipper
C/EBP	278–D K N S N E Y R V R R E R N N I A V R K S H D K A K Q R N V E T Q Q K V L E L T S D N D R L R K R V E Q L S R E L D T L R G–341
Jun	257–S Q E R I K A E R K R M R N R I A A S K C H K R K L E R I A R L E E K V K T L K A Q N S E L A S T A N M L T E Q V A Q L K O–320
Fos	233–E E R R R I R R I R R E R N K M A A A K C R N R R R E L T D T L Q A E T D Q L E D K K S A L Q T E I A N L L K E K E K L E F–296
GCN4	221–P E S S D P A A L K R A R N T E A A R R S R A R K L Q R M K O L E D K V E E L L S K N Y H L E N E V A R L K K L V G E R–COOH
YAP1	60–D L D P E T K Q K R T A Q N R A A Q R A F H E R K E R K M K E L E K K V Q S L E S I Q Q Q N E V E A T F L R D Q L I T L V N–123
CREB	279–E E A A R K R E V R L M K N R E A A R E C R R K K K E Y V K C L E N R V A V L E N Q N K T L I E E L K A L K D L Y C H K S D–342
Cys-3	95–A S R L A A E E D K R K R N T A A S A R F R I K K K Q R E Q A L E K S A K E M S E K V T Q L E G R I Q A L E T E N K Y L K G–148
CPC1	211–E D P S D V V A M K R A R N T L A A R K S B E R K A Q R L E E L E A K I E E L I A E R D R Y K N L A L A H G A S T E–COOH
HBP1	176–W D E R E L K K Q K R L S N R E S A R R S R L R K Q A E C E E L G Q R A E A L K S E N S S L R I E L D R I K K E Y E E L L S–239
TGA1	68–S K P V E K V L R R L A Q R N E A A R K S R L R K K A Y V Q Q L E N S K L K L I Q L E Q E L E R A R K Q G M C V G G G V D A–131
Opaque2	223–M P T E E R V R K R K E S N R E S A R R S R Y R K A A H L K E L E D Q V A Q L K A E N S C L L R R I A A L N Q K Y N D A N V–286

Figure 10.27 Amino acid sequence comparisons of several DNA binding proteins

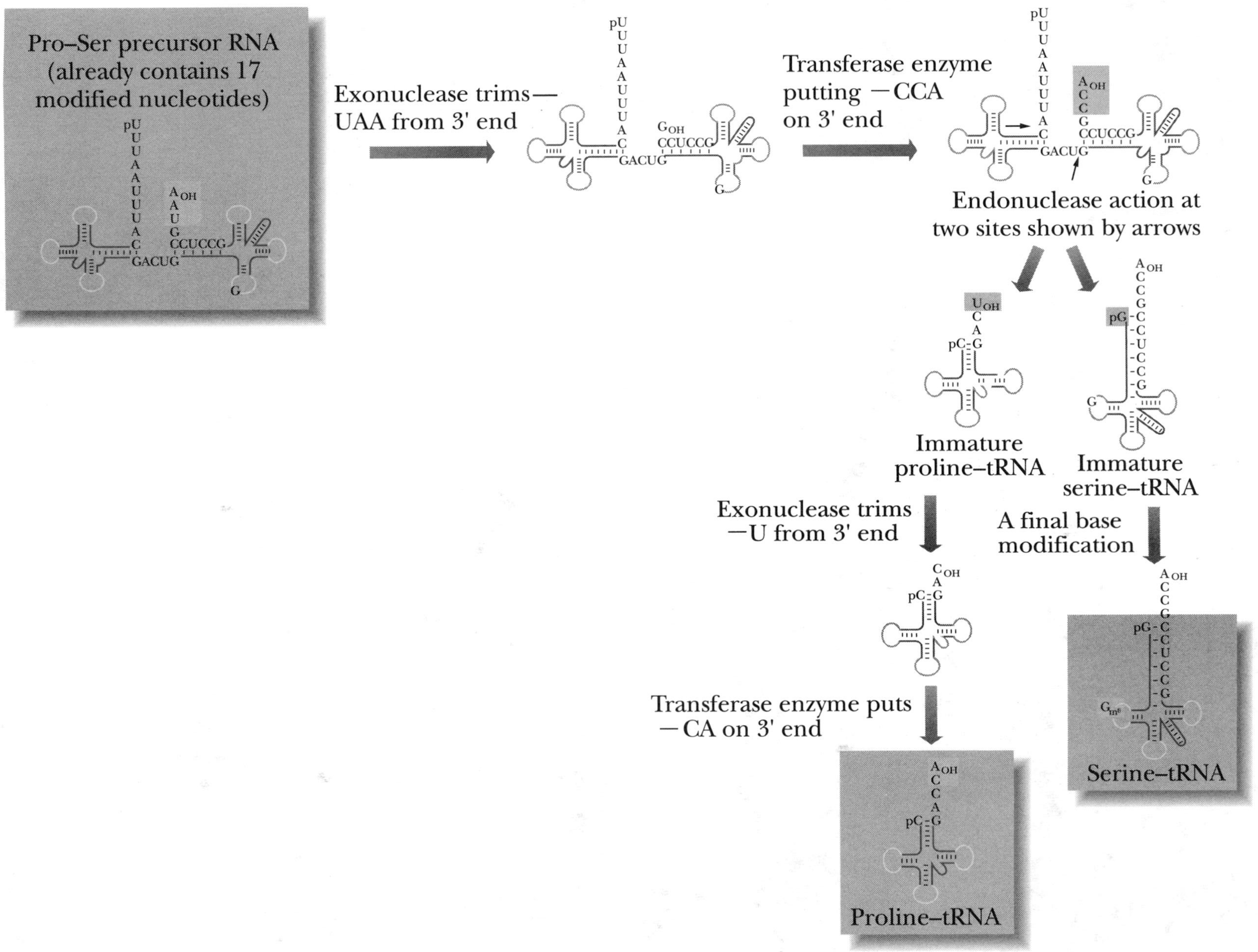

Figure 10.30 Posttranscriptional modification of tRNA precursor

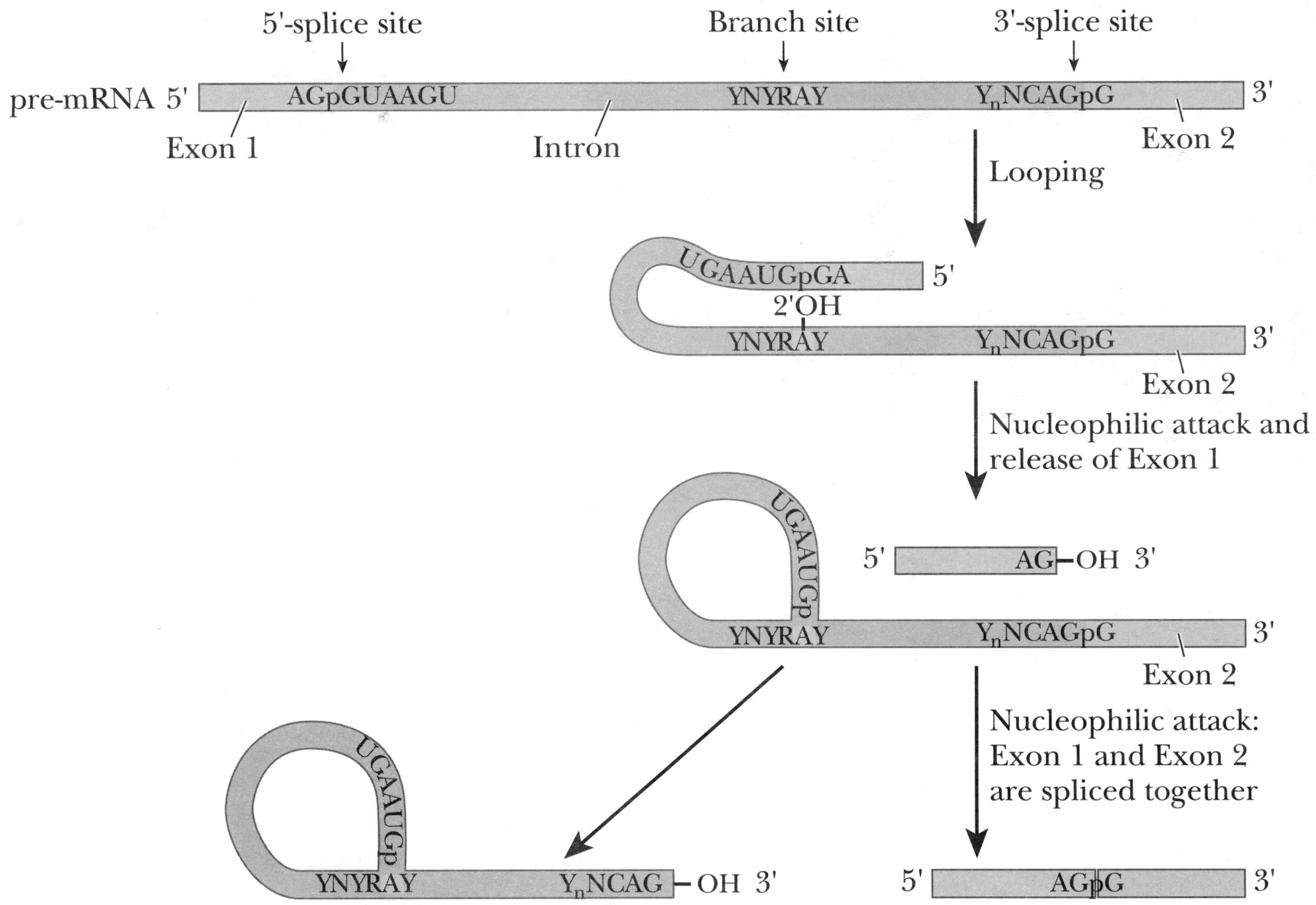

5'-splice site
Branch site
3'-splice site
pre-mRNA 5'
AGpGUAAGU
YNYRAY
Y_nNCAGpG
3'
Exon 1
Intron
Exon 2
Looping
UGAAUGpGA
5'
2'OH
YNYRAY
Y_nNCAGpG
3'
Exon 2
Nucleophilic attack and release of Exon 1
UGAAUGp
5'
AG—OH 3'
YNYRAY
Y_nNCAGpG
3'
Exon 2
Nucleophilic attack: Exon 1 and Exon 2 are spliced together
UGAAUGp
YNYRAY
Y_nNCAG—OH 3'
5'
AGpG
3'

Figure 10.34 Splicing of mRNA precursors

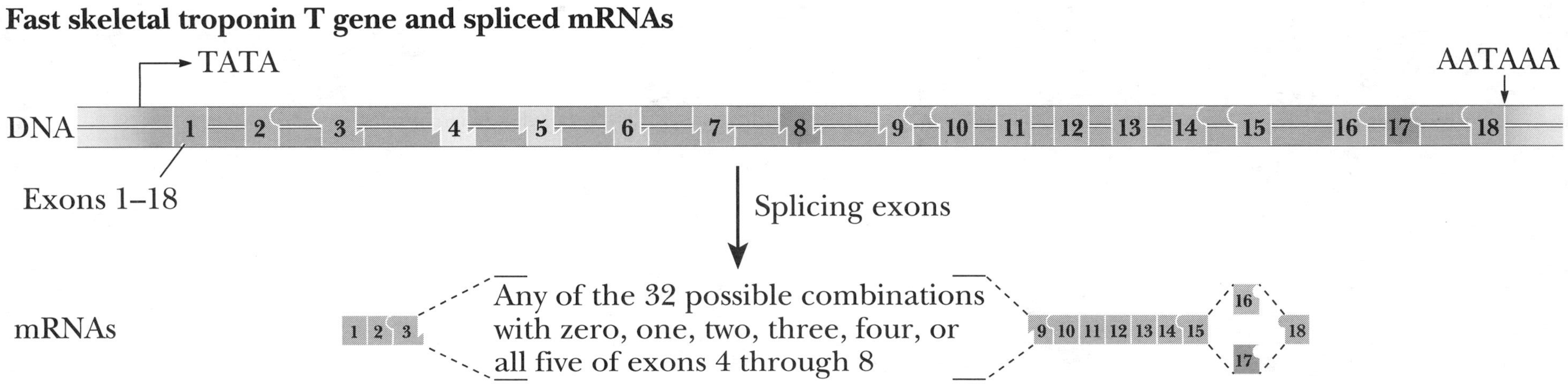

Figure 10.35 Organization of the fast skeletal muscle troponin T gene

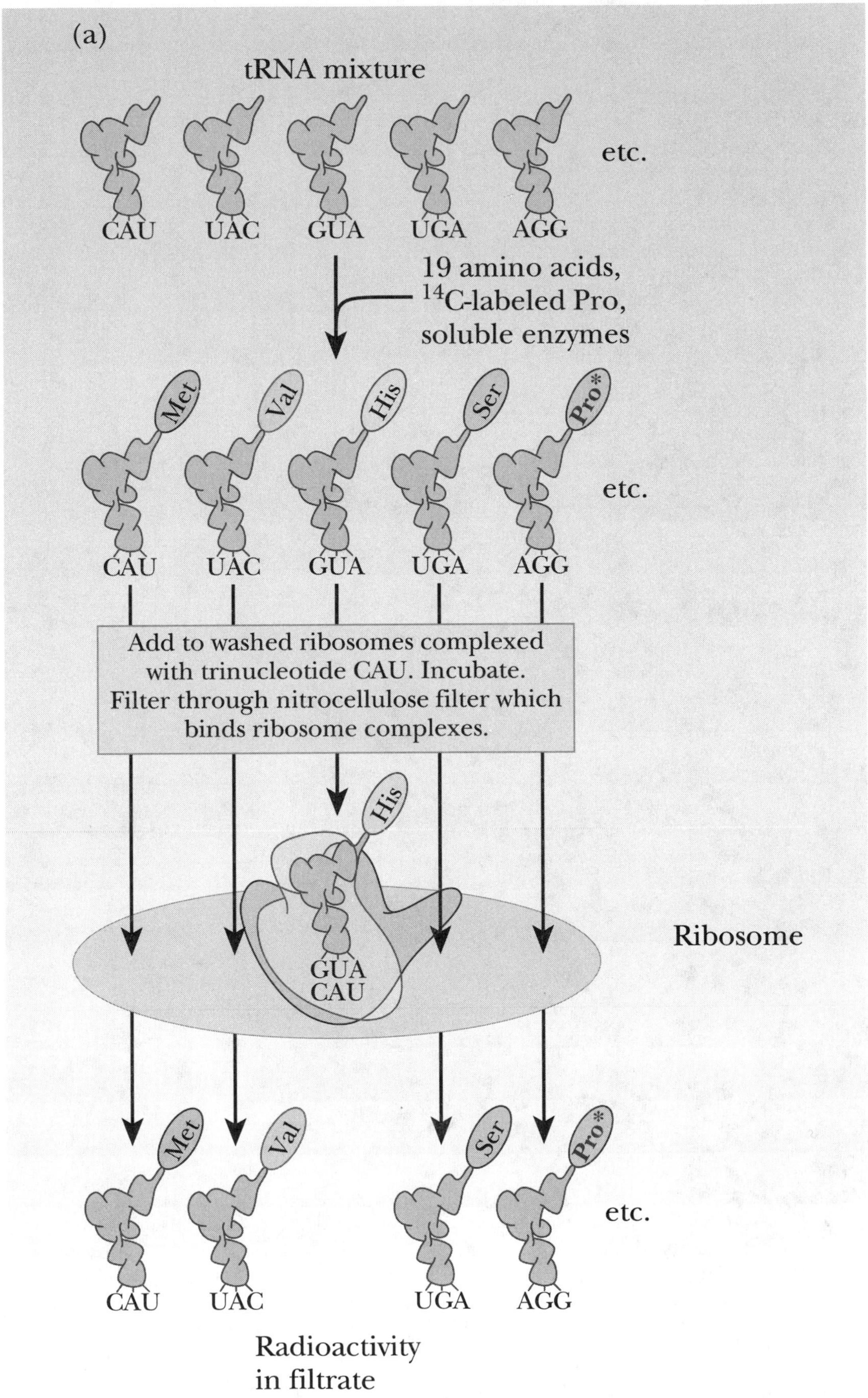

Figure 11.3a The filter-binding assay for elucidation of the genetic code

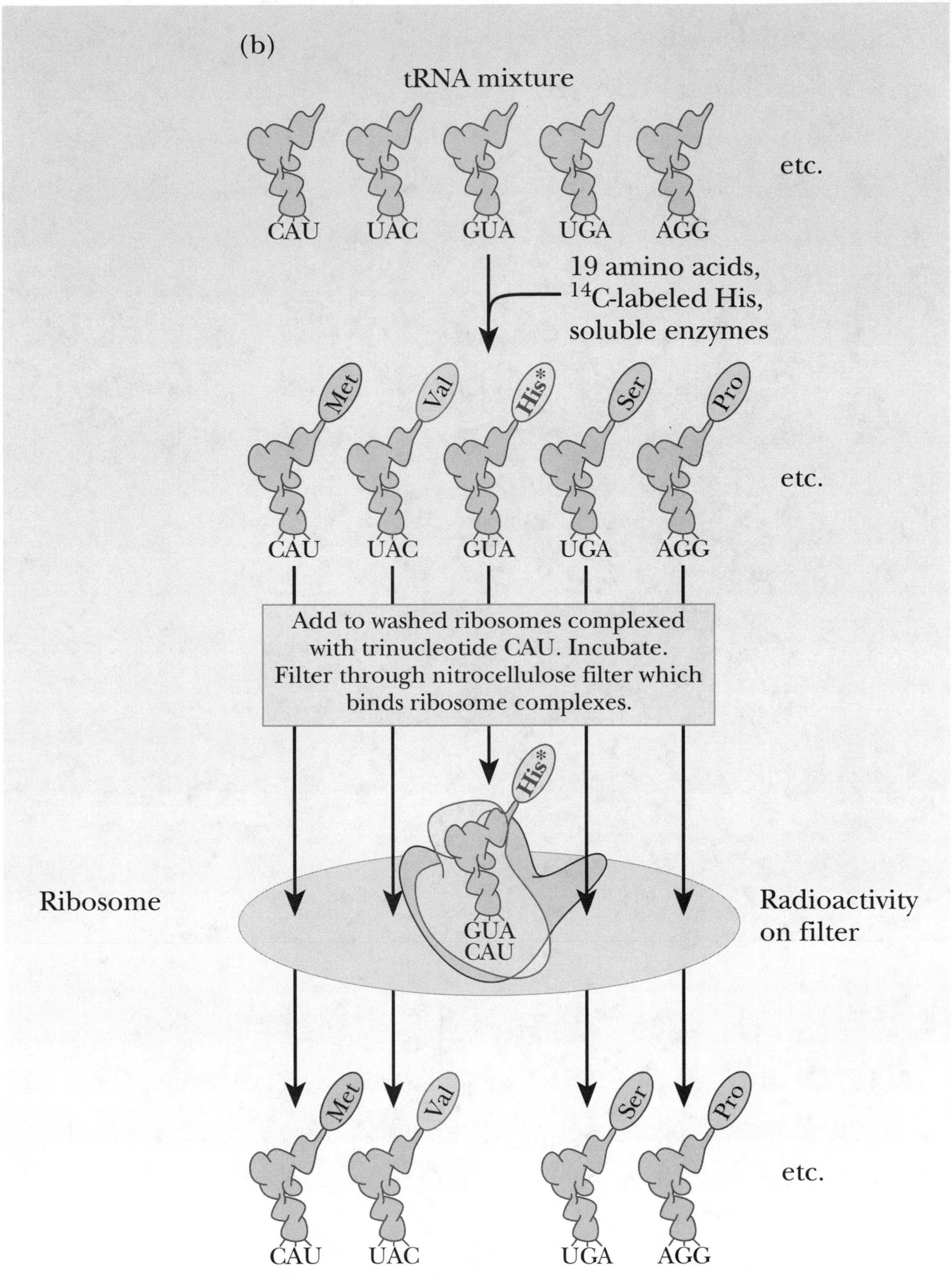

Figure 11.3b The filter-binding assay for elucidation of the genetic code

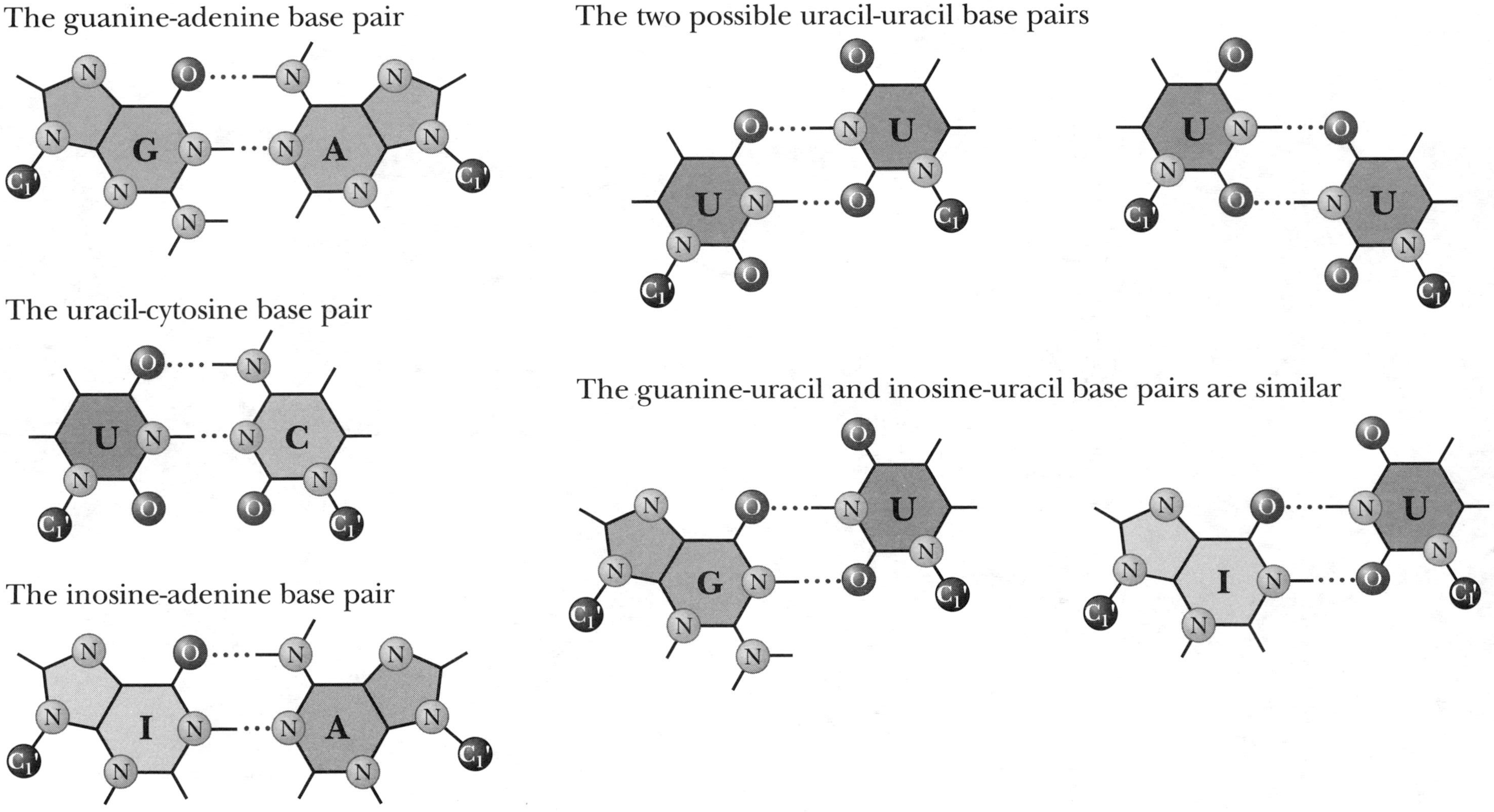

Figure 11.5 Various base-pairing alternatives

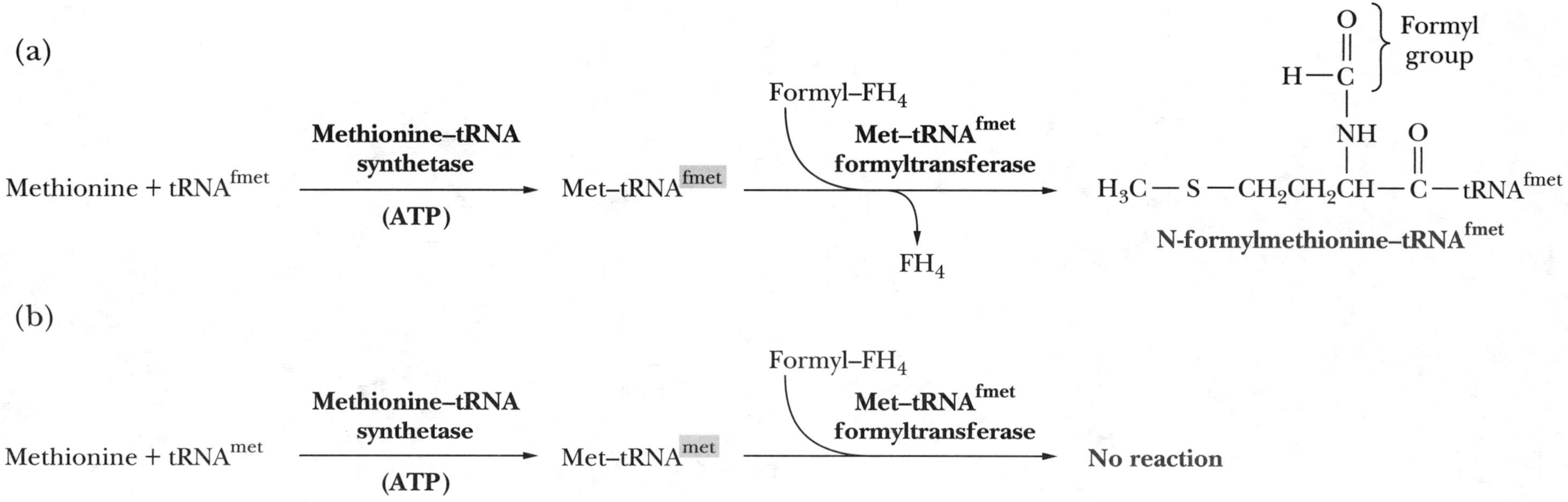

Figure 11.9 Formation of the N-formylmethionine-tRNA

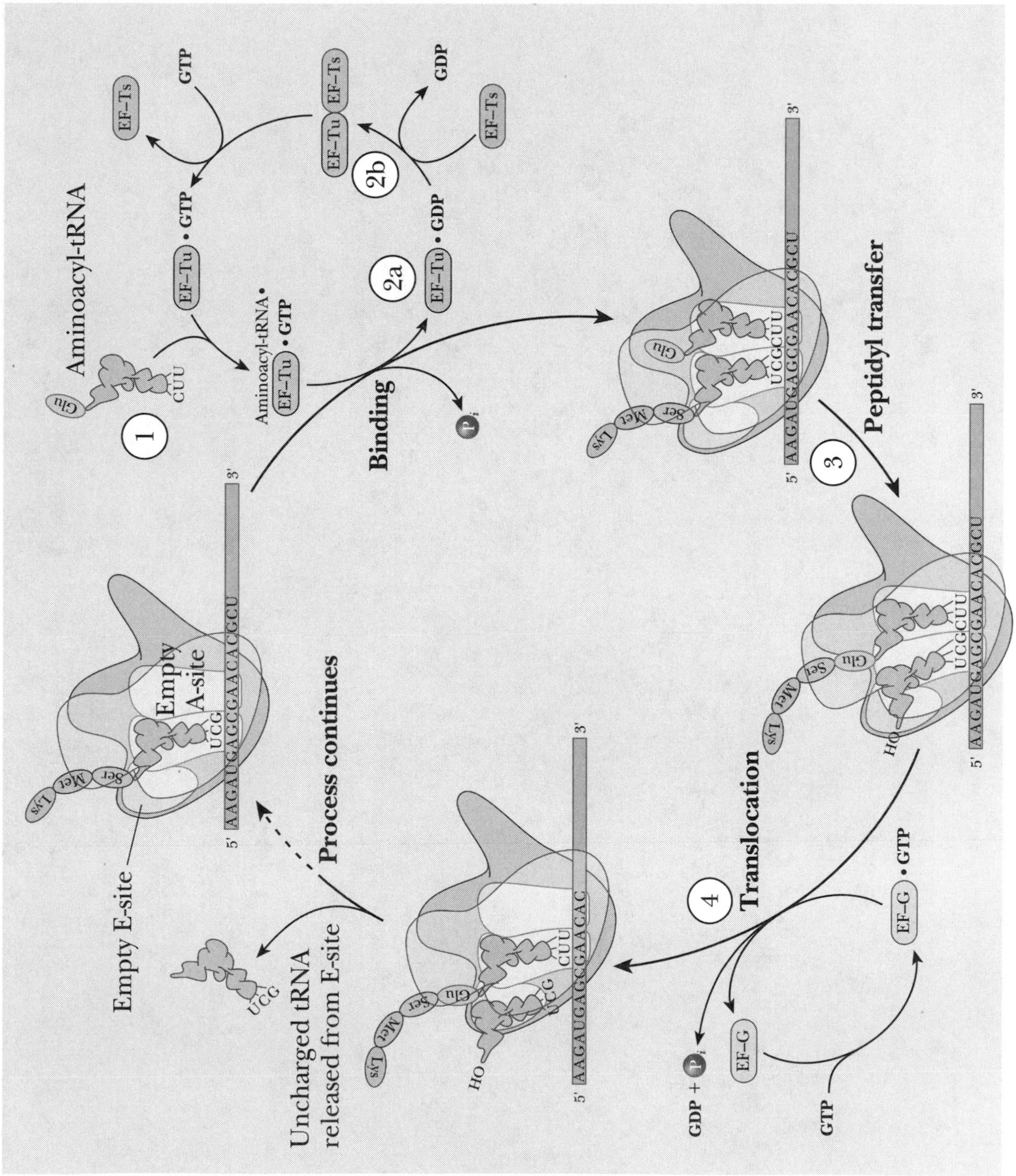

Figure 11.12 A summary of the steps in chain elongation

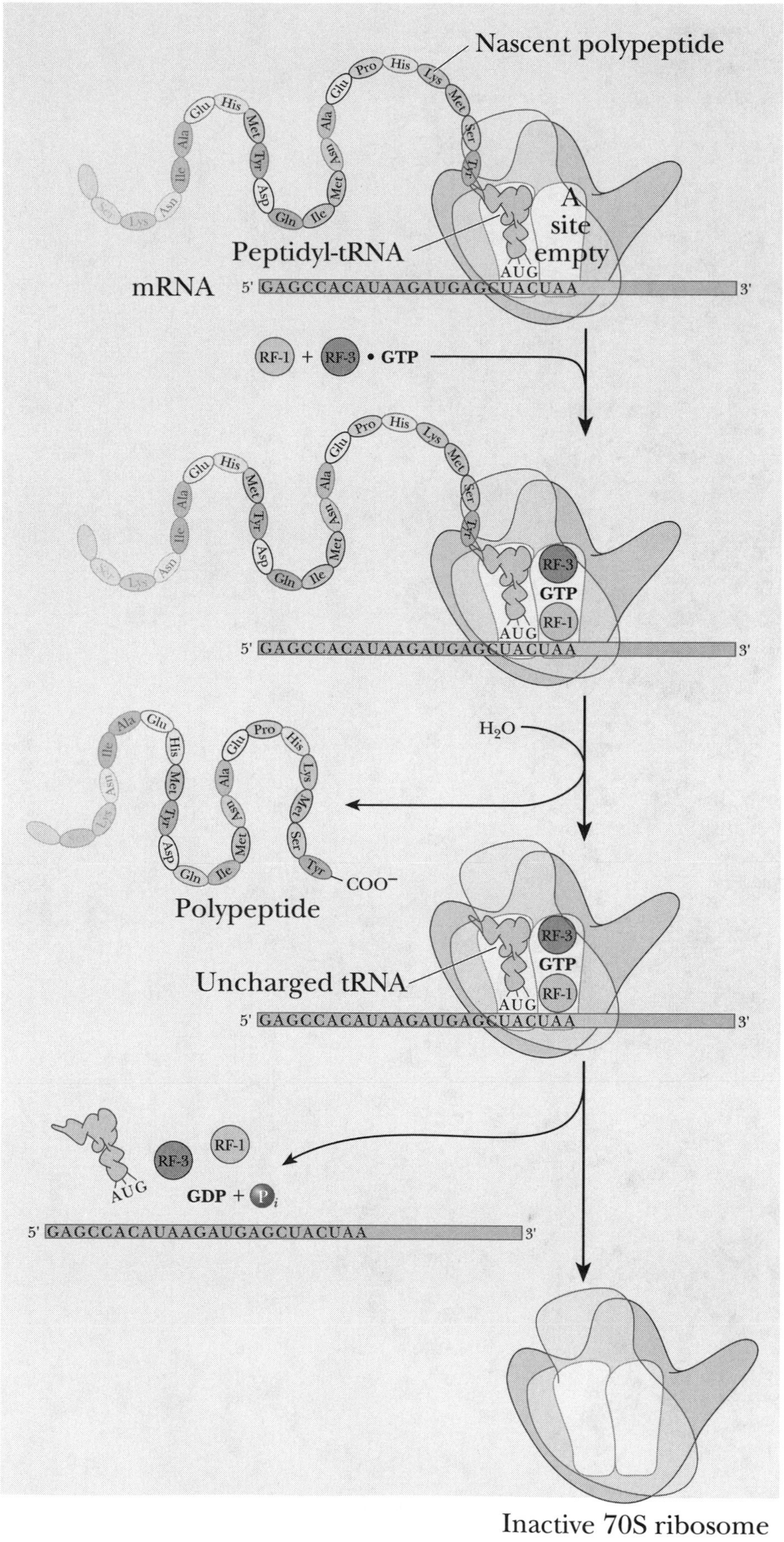

Figure 11.15 The events in peptide chain termination

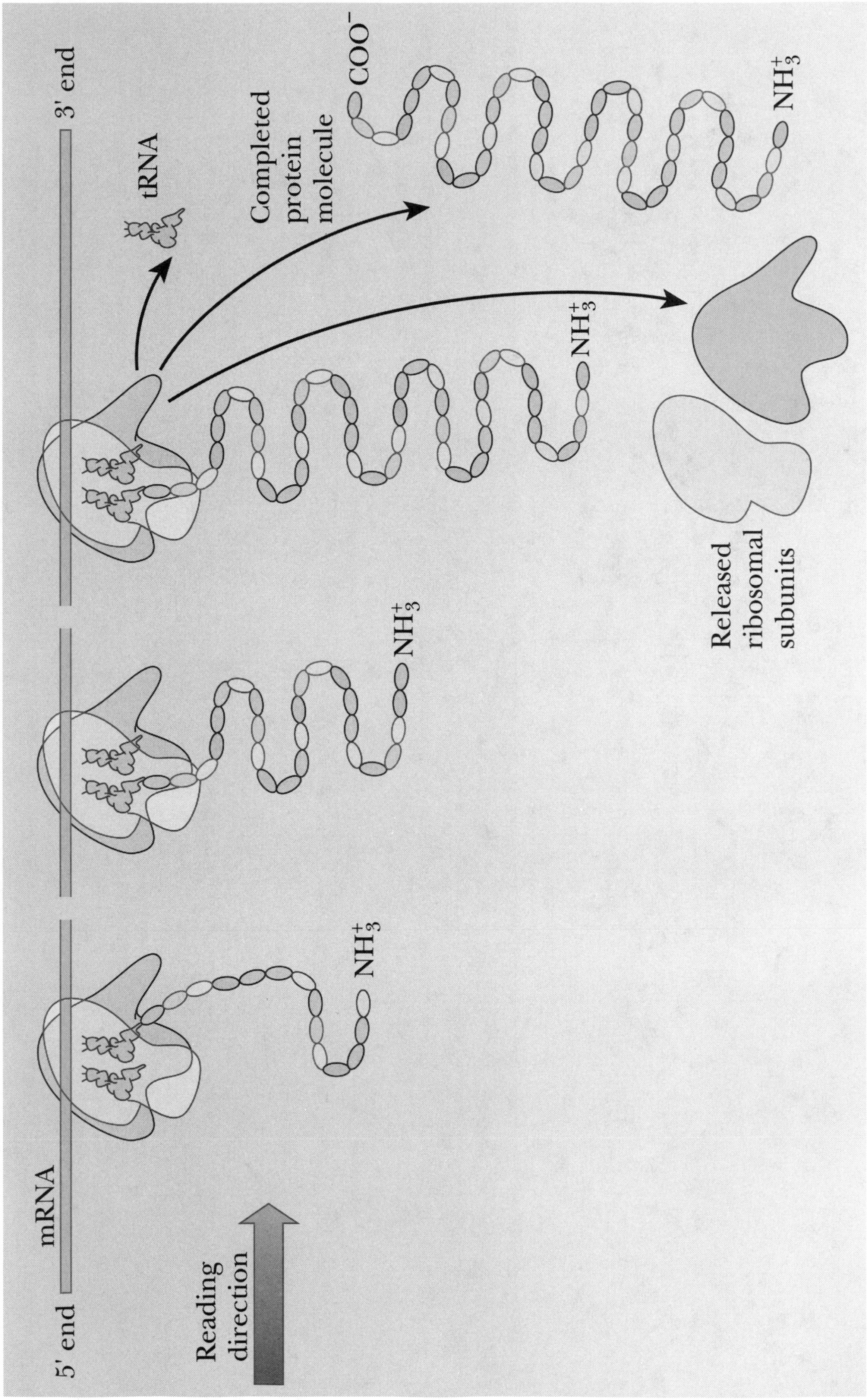

Figure 11.17 Simultaneous protein synthesis on polysomes

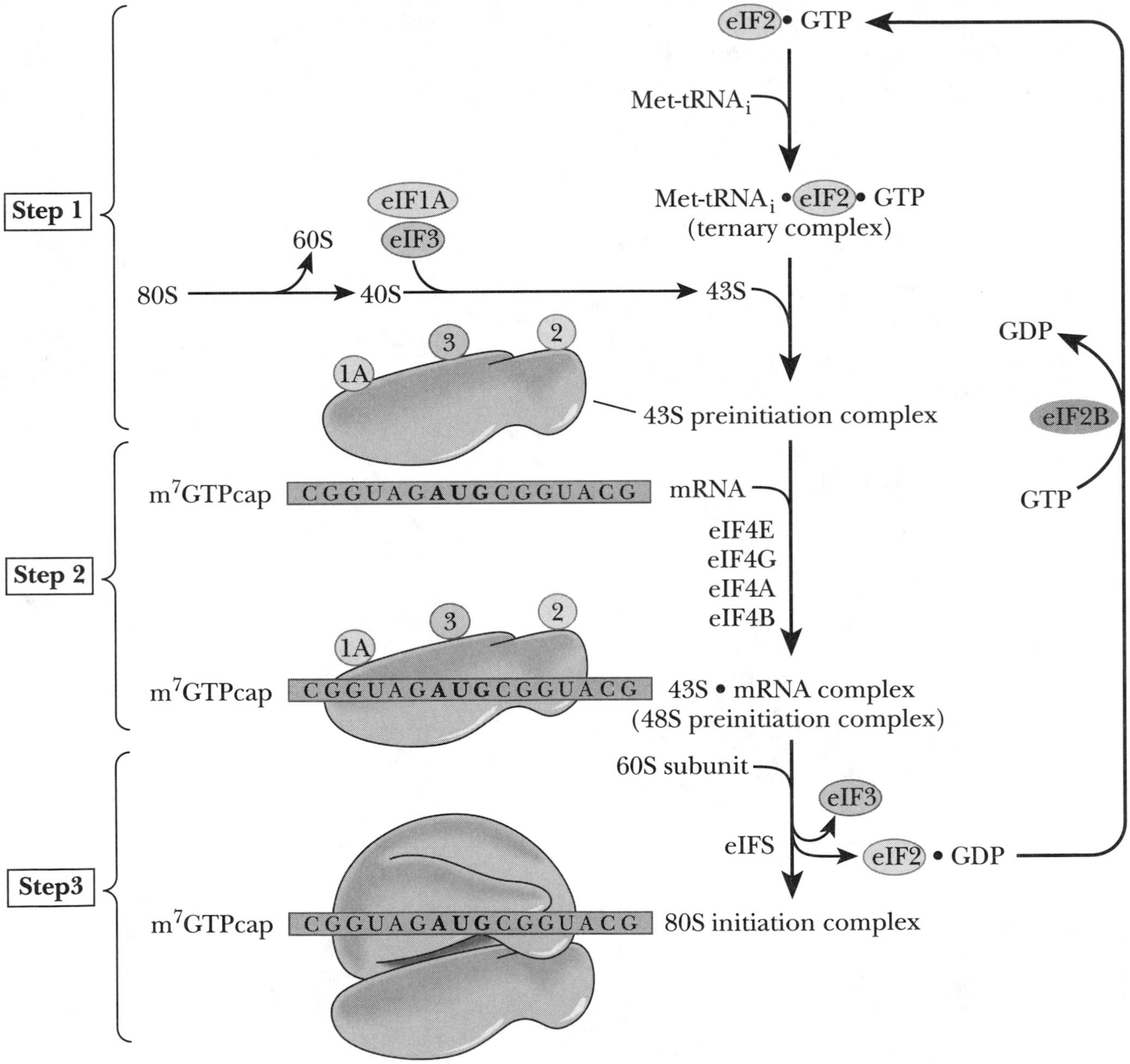

Figure 11.20 The three stages in the initiation of translation in eukaryotic cells

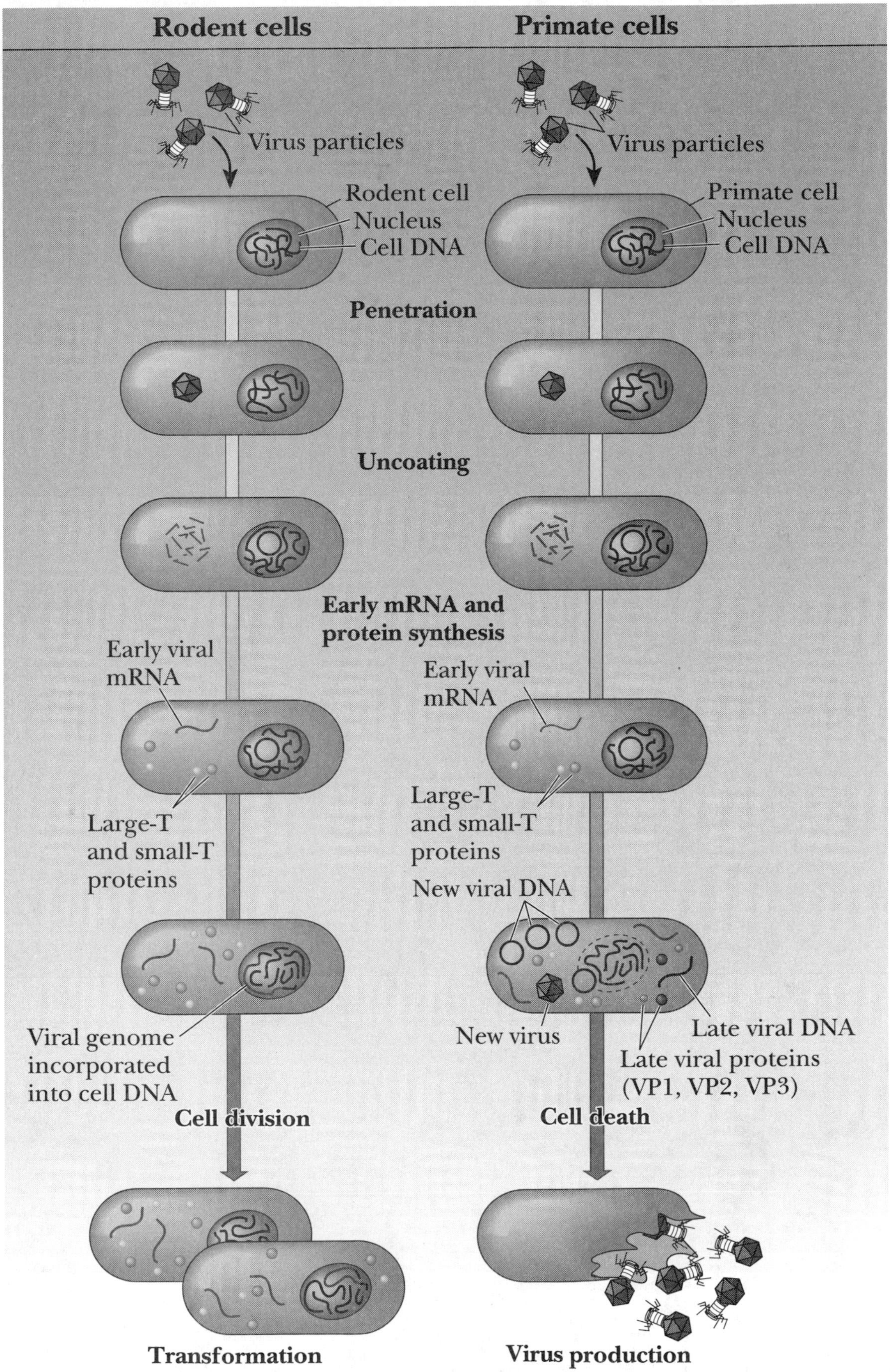

Figure 11.25 The outcome of infection of cells by simian viris 40 depends on the nature of the cells

TABLE 11.1 The Genetic Code

First Position (5′-end)	Second Position U	Second Position C	Second Position A	Second Position G	Third Position (3′-end)
U	UUU Phe	UCU Ser	UAU Tyr	UGU Cys	U
	UUC Phe	UCC Ser	UAC Tyr	UGC Cys	C
	UUA Leu	UCA Ser	UAA Stop	UGA Stop	A
	UUG Leu	UCG Ser	UAG Stop	UGG Trp	G
C	CUU Leu	CCU Pro	CAU His	CGU Arg	U
	CUC Leu	CCC Pro	CAC His	CGC Arg	C
	CUA Leu	CCA Pro	CAA Gln	CGA Arg	A
	CUG Leu	CCG Pro	CAG Gln	CGG Arg	G
A	AUU Ile	ACU Thr	AAU Asn	AGU Ser	U
	AUC Ile	ACC Thr	AAC Asn	AGC Ser	C
	AUA Ile	ACA Thr	AAA Lys	AGA Arg	A
	AUG Met*	ACG Thr	AAG Lys	AGG Arg	G
G	GUU Val	GCU Ala	GAU Asp	GGU Gly	U
	GUC Val	GCC Ala	GAC Asp	GGC Gly	C
	GUA Val	GCA Ala	GAA Glu	GGA Gly	A
	GUG Val	GCG Ala	GAG Glu	GGG Gly	G

*AUG signals translation initiation as well as coding for Met residues.

Third-Base Degeneracy Is Color-Coded

Third-Base Relationship	Third Bases with Same Meaning	Number of Codons
Third-base irrelevant	U, C, A, G	32 (8 families)
Purines	A or G	12 (6 pairs)
Pyrimidines	U or C	14 (7 pairs)
Three out of four	U, C, A	3 (AUX = Ile)
Unique definitions	G only	2 (AUG = Met) (UGG = Trp)
Unique definition	A only	1 (UGA = Stop)

Table 11.1 The genetic code

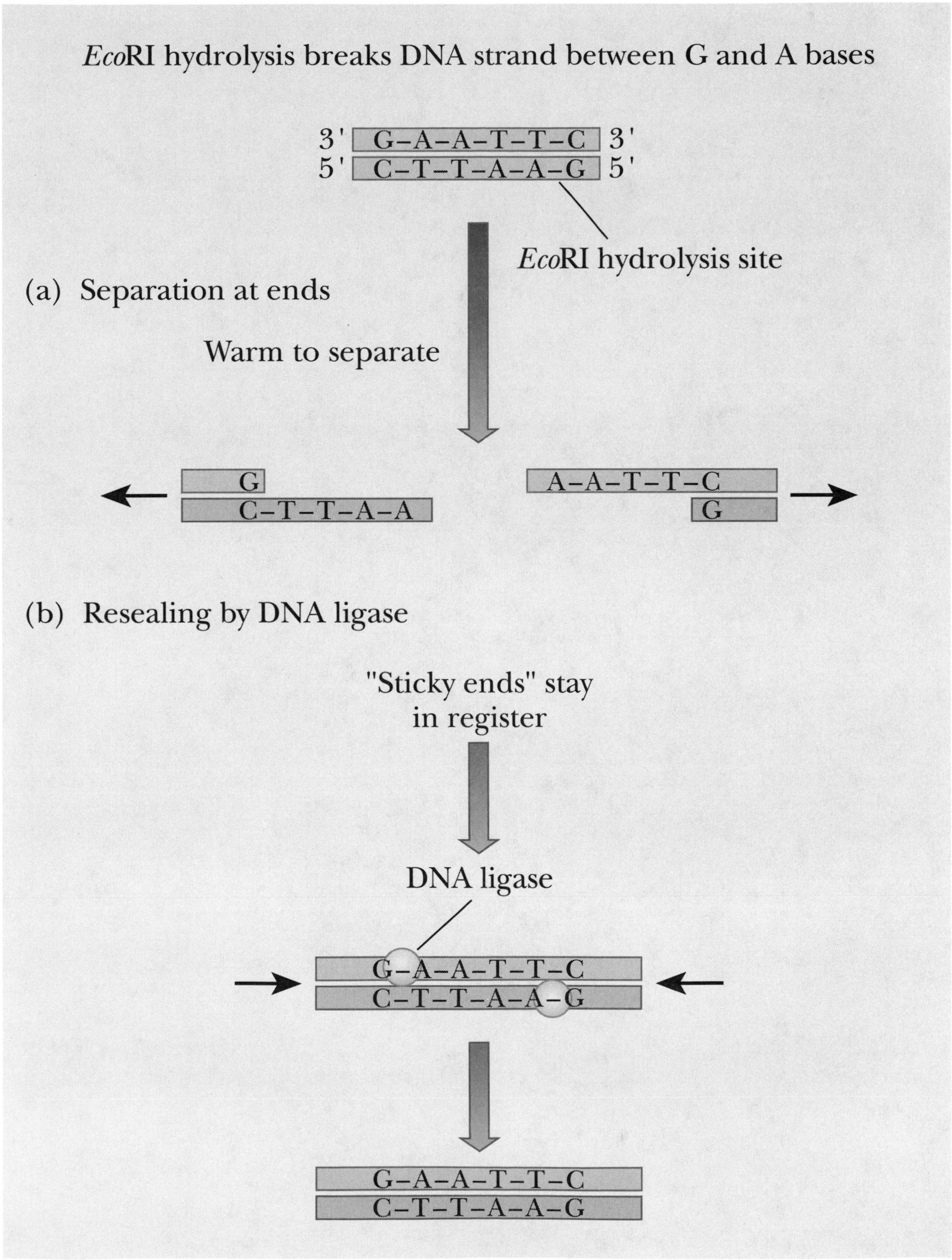

Figure B.4 Hydrolysis of DNA by restriction endonucleases

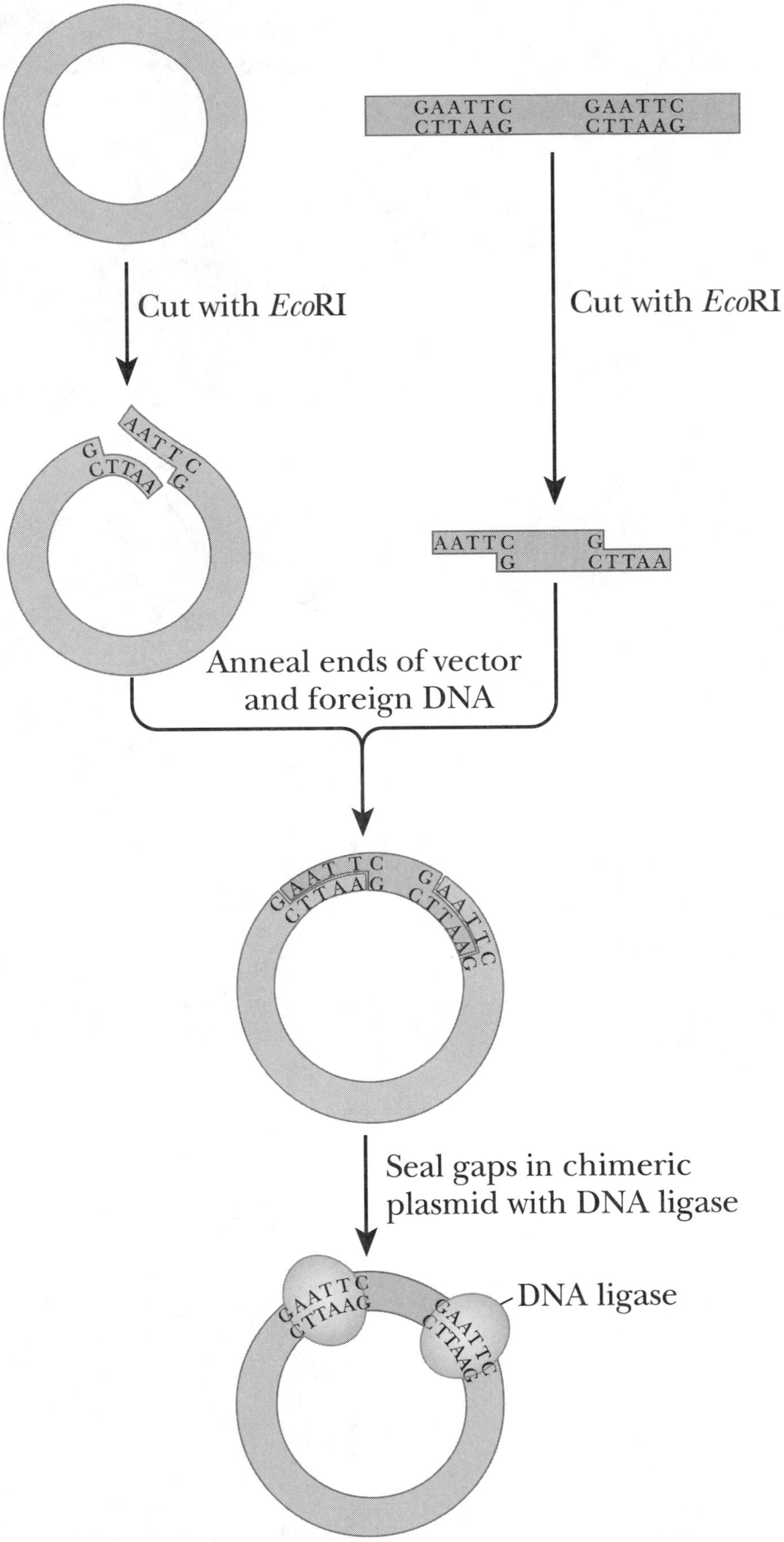

Figure B.5 Restriction endonuclease EcoRI cleaves double-stranded DNA

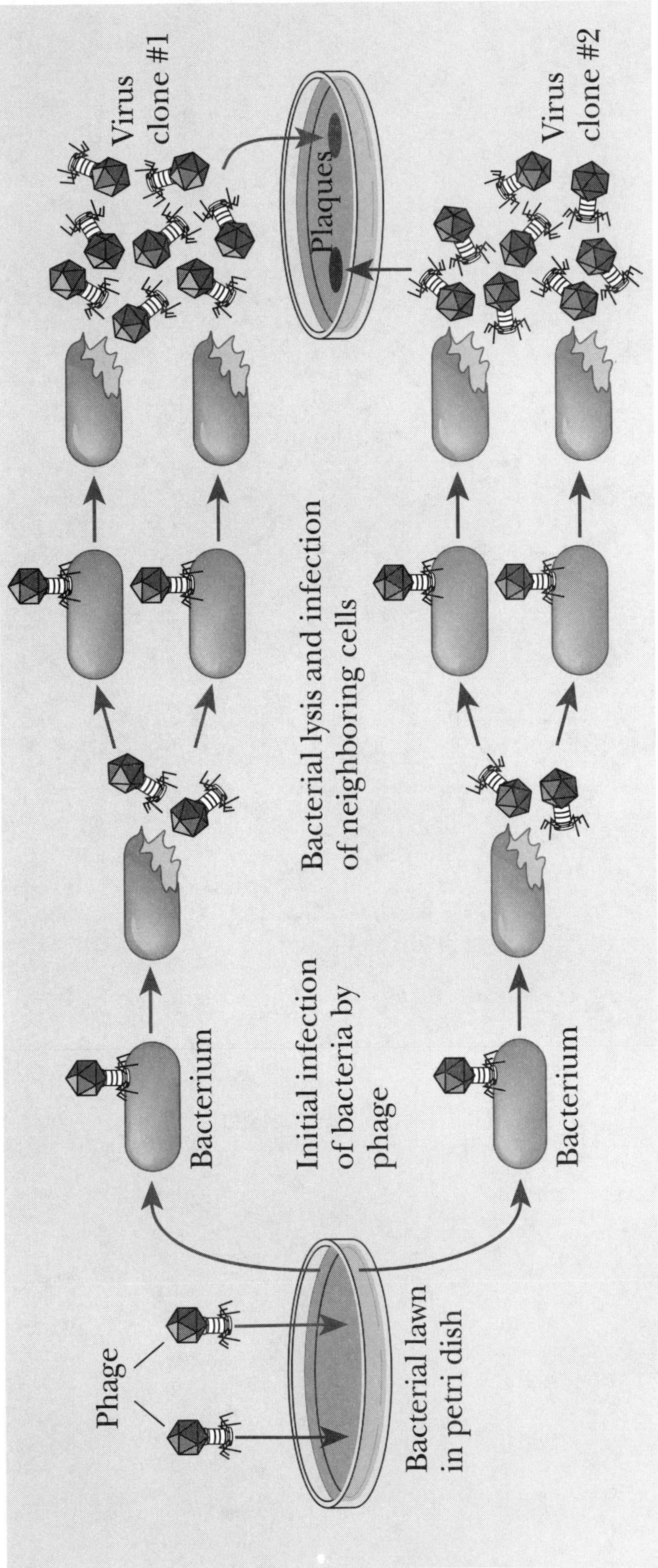

(*Adapted from* Dealing with Genes: The Language of Heredity, *by Paul Berg and Maxine Singer,* © *1992 by University Science Books*)

Figure B.6 The cloning of a virus

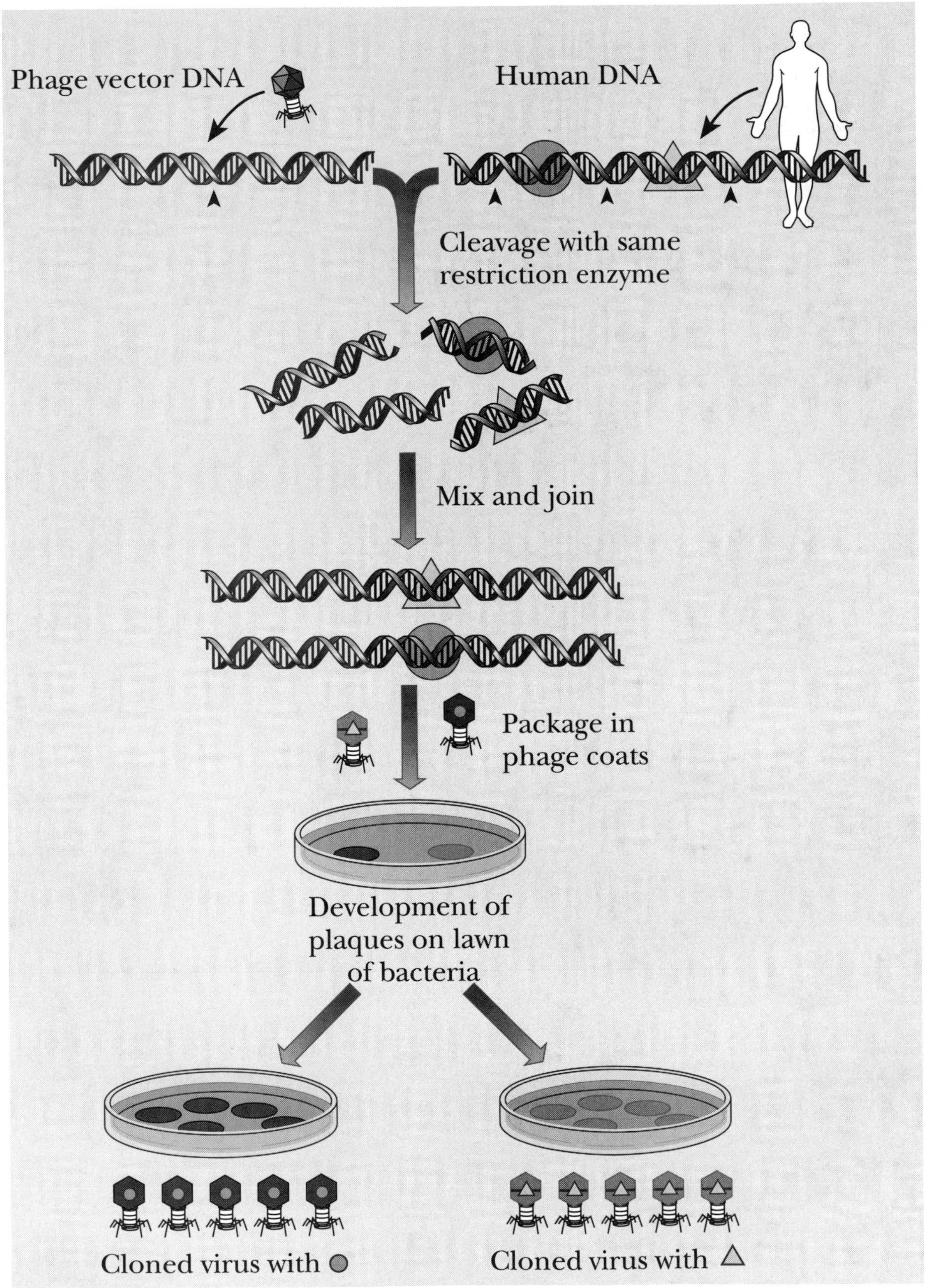

(Adapted from Dealing with Genes: The Language of Heredity, *by Paul Berg and Maxine Singer, © 1992 by University Science Books)*

Figure B.8 The cloning of human DNA fragments with a viral vector

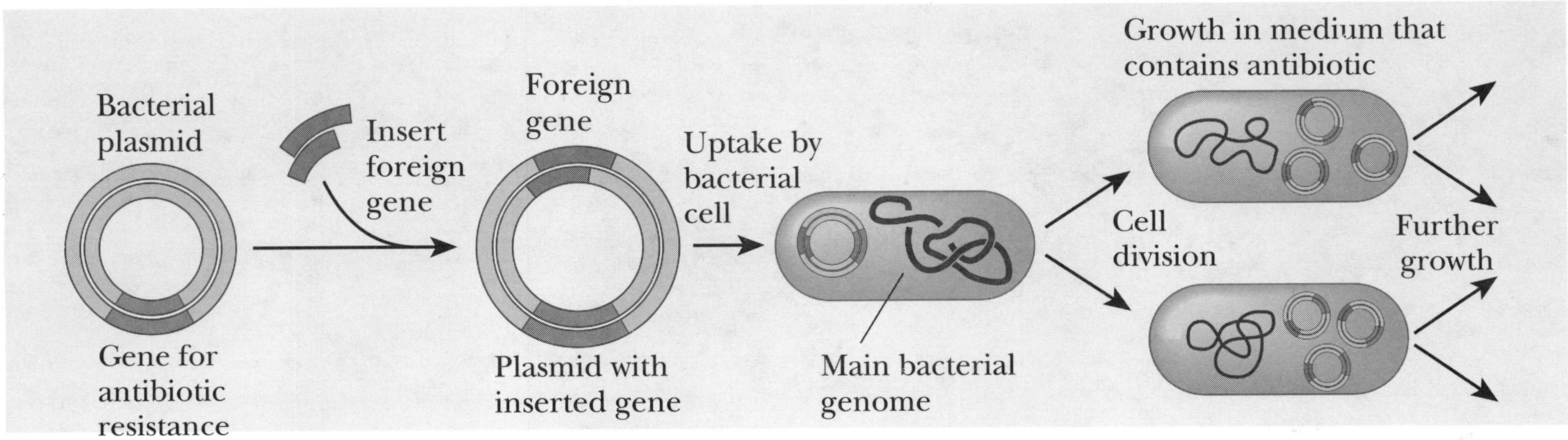

(*Adapted from* Dealing with Genes: The Language of Heredity, *by Paul Berg and Maxine Singer, © 1992 by University Science Books*)

Figure B.9 Selecting for recombinant DNA in a bacterial plasmid

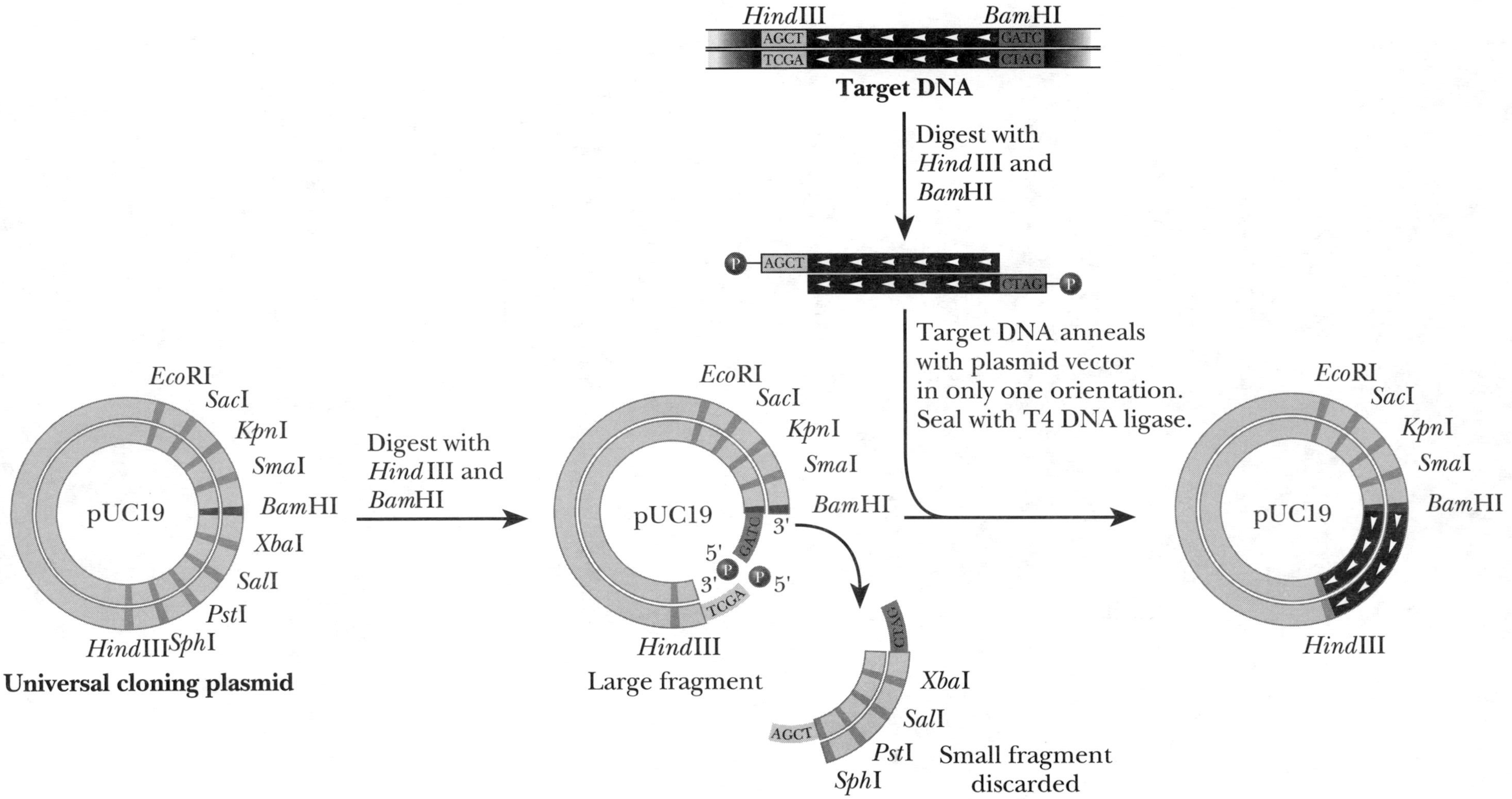

HindIII
BamHI
AGCT
TCGA
GATC
CTAG
Target DNA
Digest with HindIII and BamHI
P
Target DNA anneals with plasmid vector in only one orientation. Seal with T4 DNA ligase.
pUC19
EcoRI
SacI
KpnI
SmaI
BamHI
XbaI
SalI
PstI
SphI
HindIII
Universal cloning plasmid
Digest with HindIII and BamHI
3'
5'
Large fragment
Small fragment discarded

Figure B.12 The pUC series of plasmids

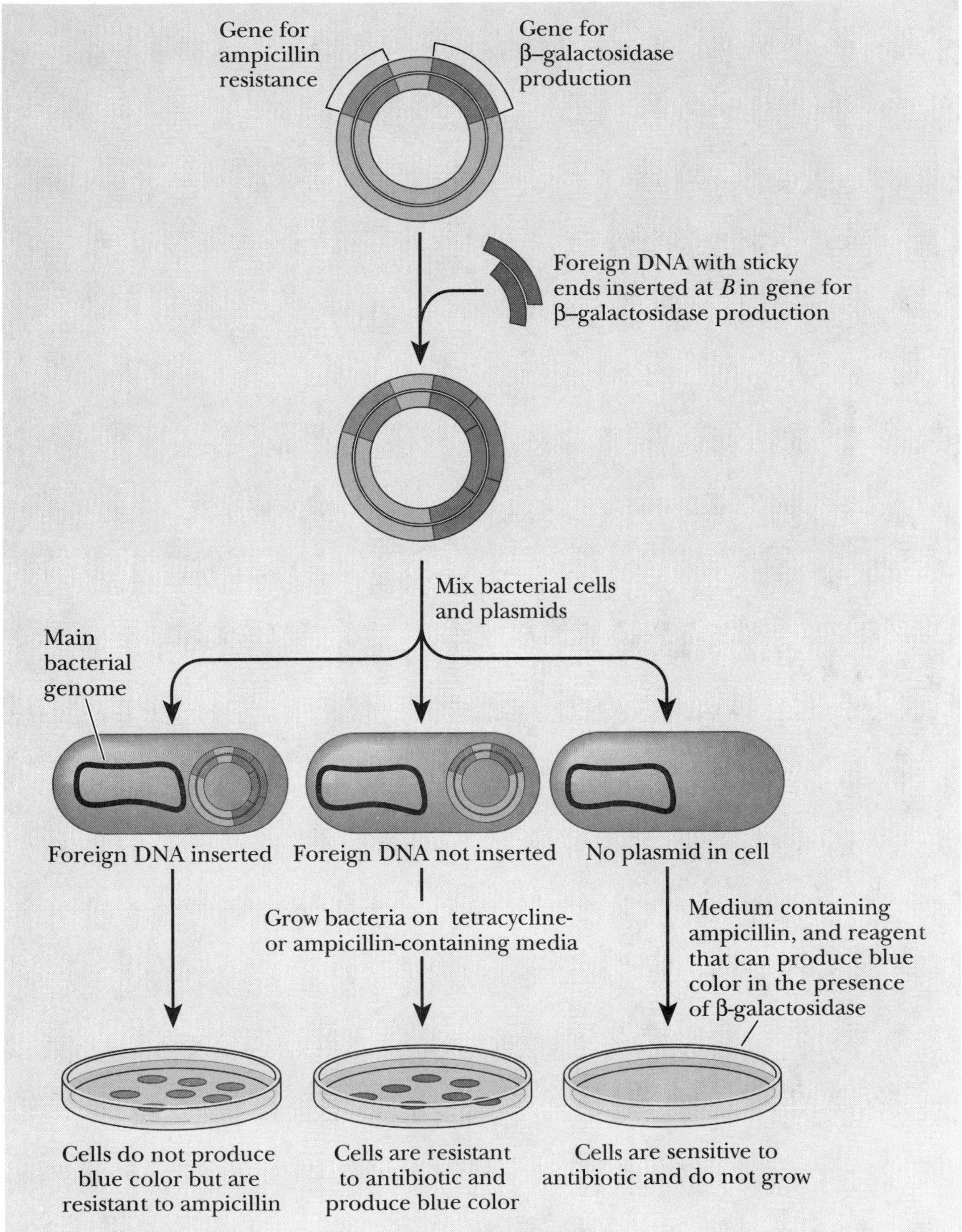

(*Adapted from* Dealing with Genes: The Language of Heredity, *by Paul Berg and Maxine Singer, © 1992 by University Science Books)*

Figure B.13 Clone selection via blue/white screening

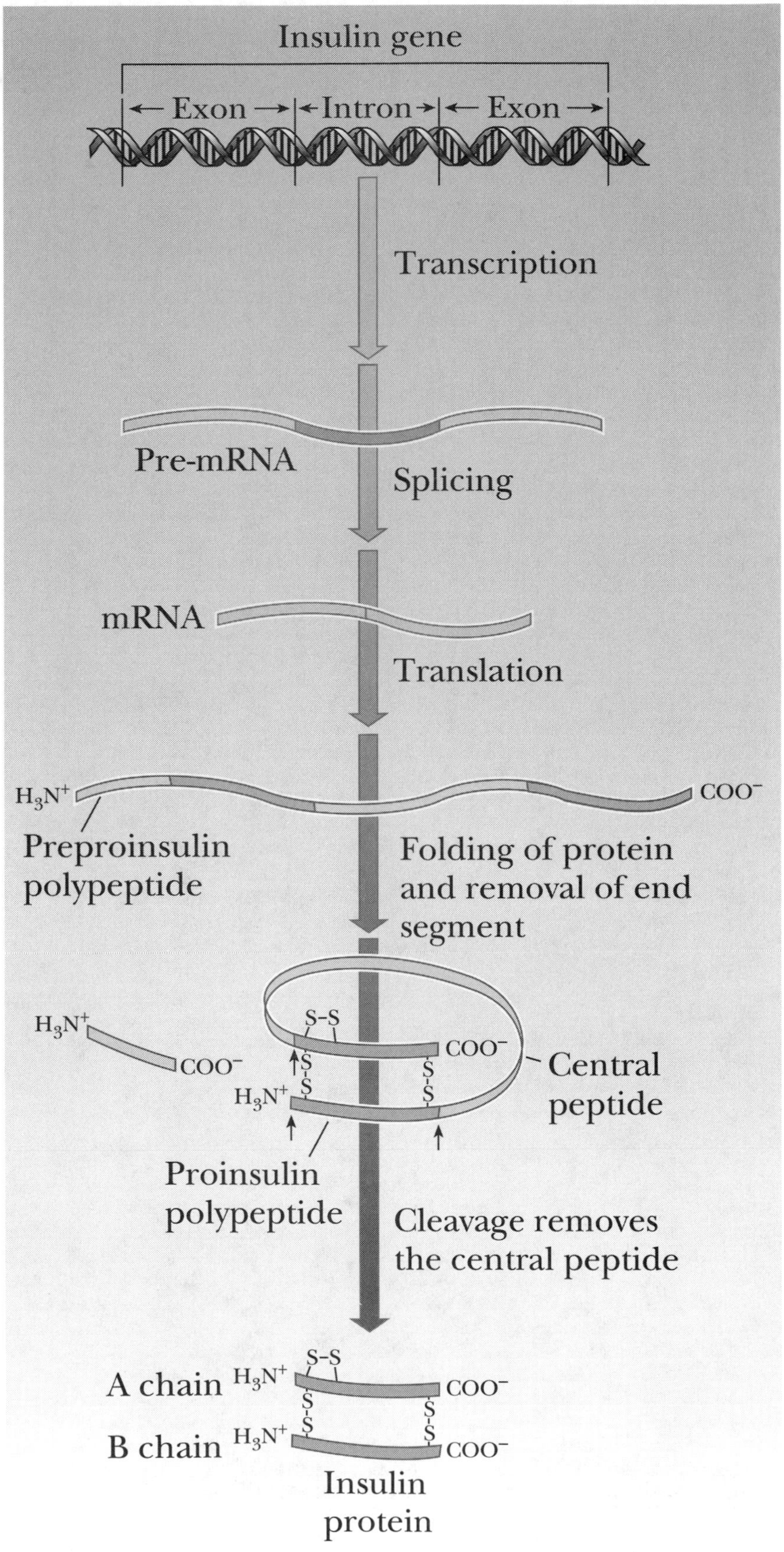

(Adapted from Dealing with Genes: The Language of Heredity, *by Paul Berg and Maxine Singer, © 1992 by University Science Books)*

Figure B.15 Synthesis of insulin in humans

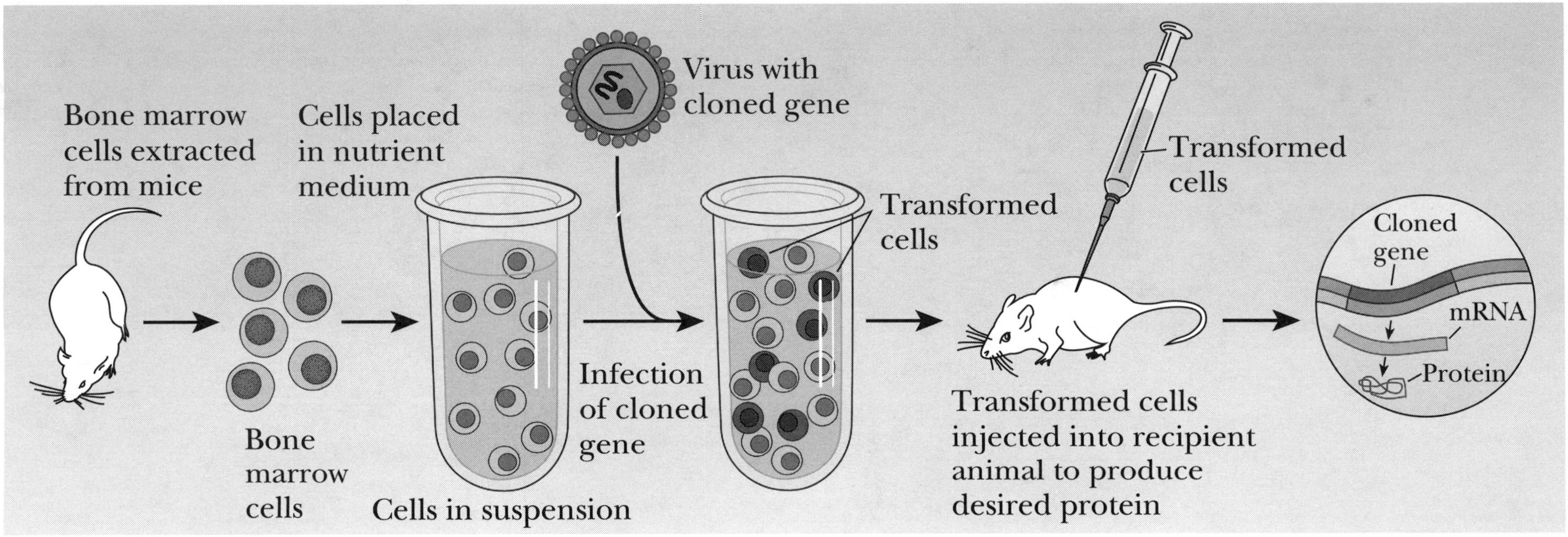

(*Adapted from* Dealing with Genes: The Language of Heredity, *by Paul Berg and Maxine Singer, © 1992 by University Science Books.)*

Figure B.18 Gene therapy in bone marrow cells

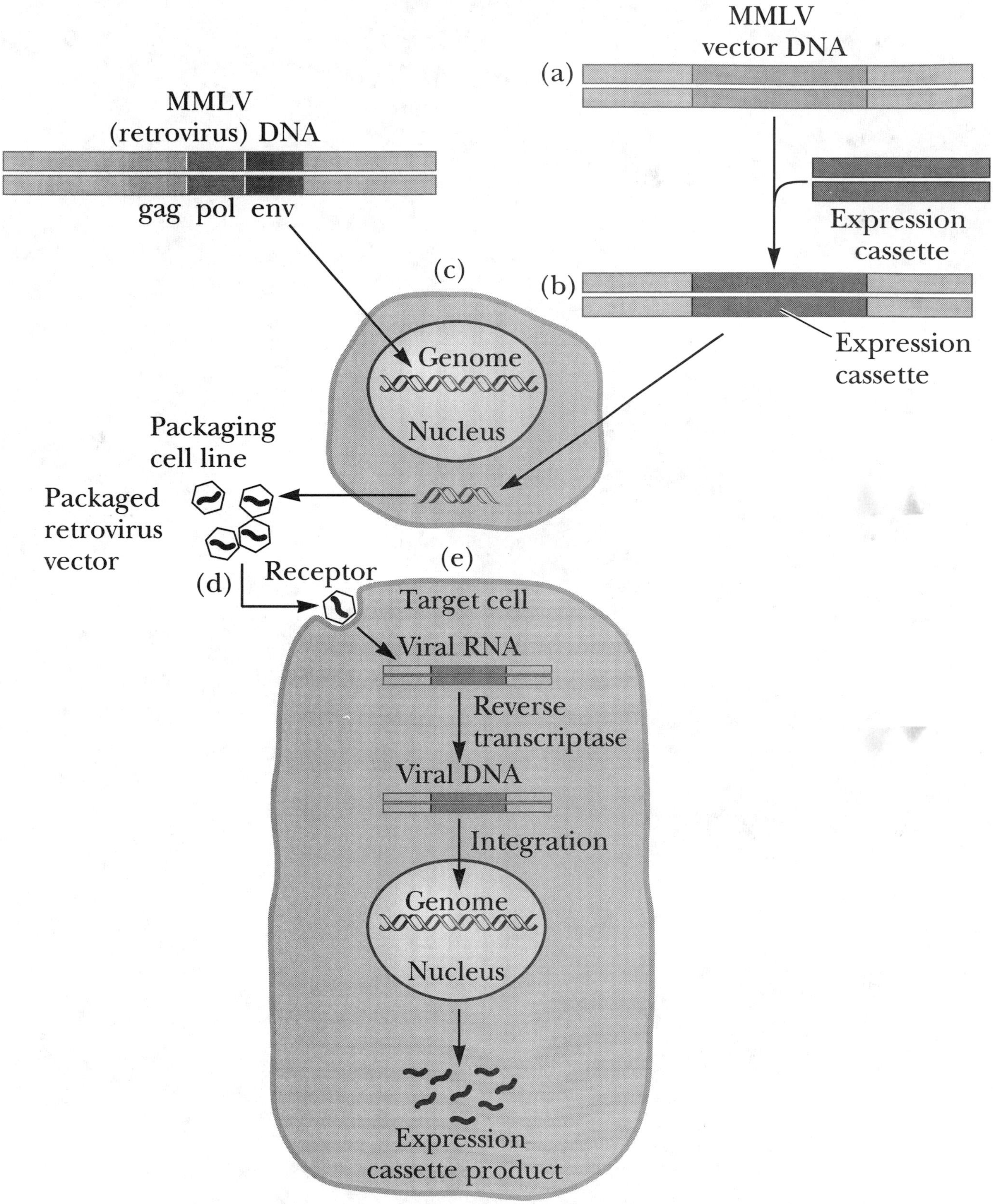

(Adapted from Figure 1 in Crystal, R. G., 1995. Transfer of genes to humans: Early lessons and obstacles to success. Science ***270:*** *404.)*

Figure B.19 Human gene therapy via retroviruses

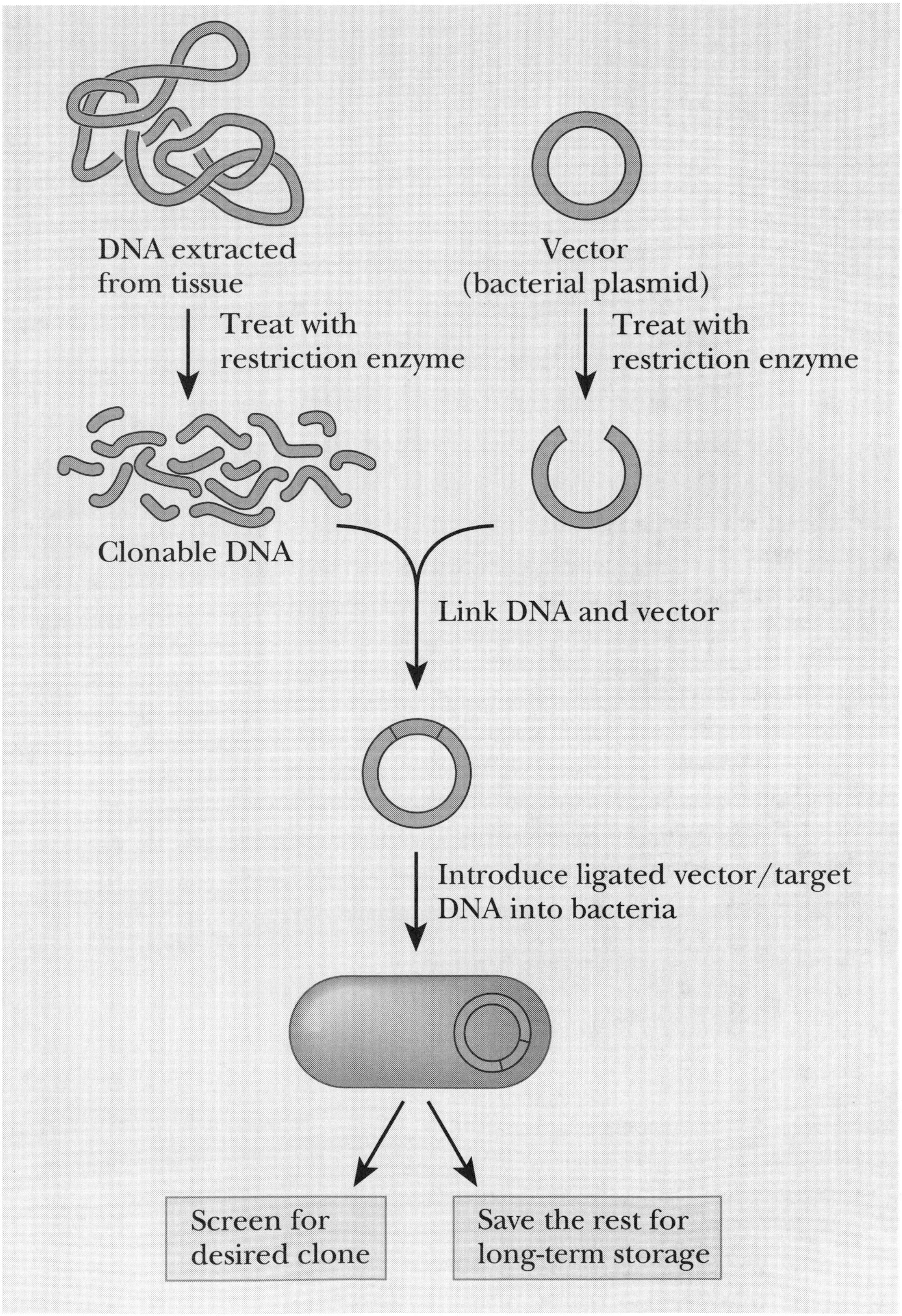

(Adapted from Dealing with Genes: The Language of Heredity, *by Paul Berg and Maxine Singer, © 1992 by University Science Books)*

Figure B.23 Steps involved in the construction of a DNA library

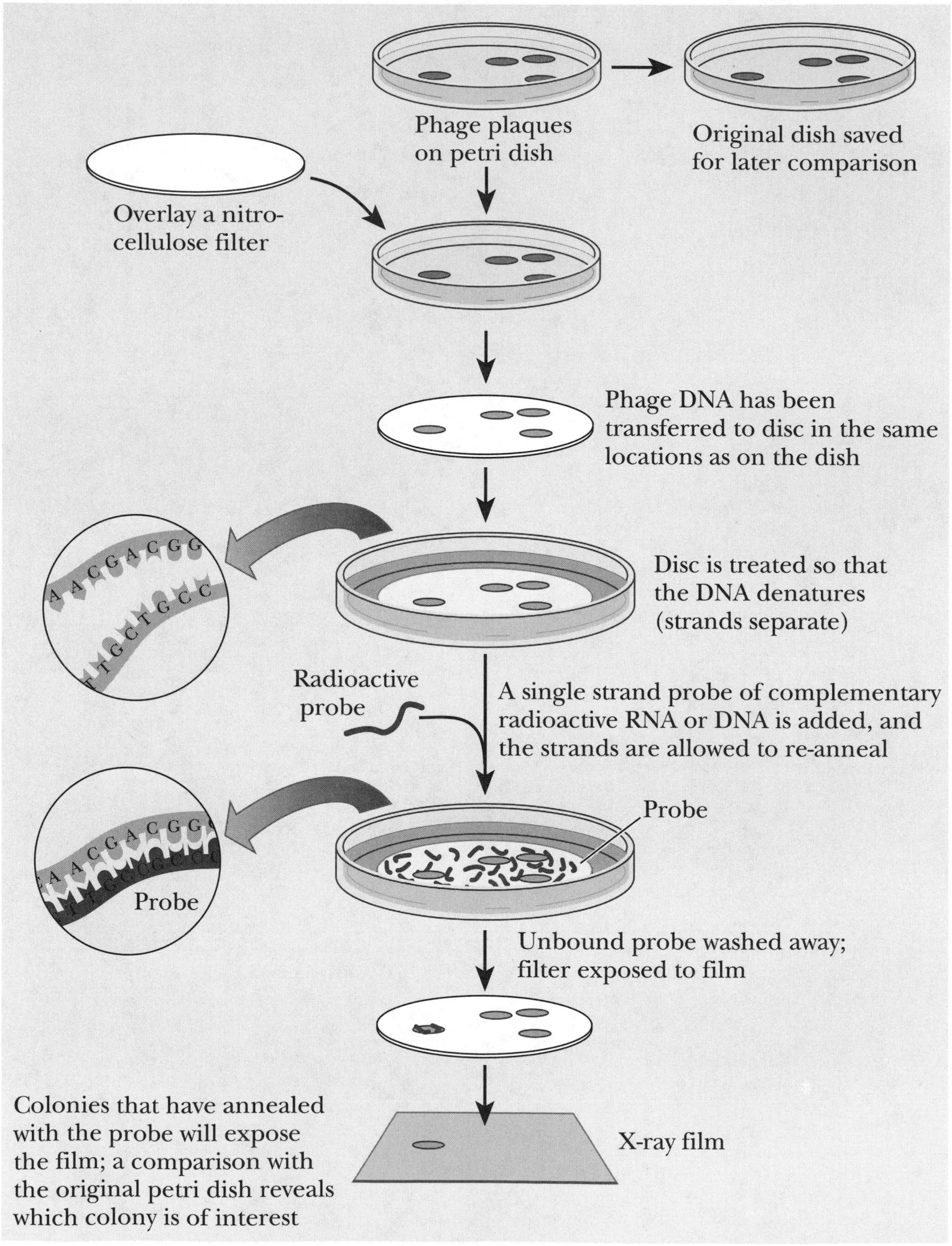

Phage plaques on petri dish
Original dish saved for later comparison
Overlay a nitro-cellulose filter
Phage DNA has been transferred to disc in the same locations as on the dish
Disc is treated so that the DNA denatures (strands separate)
Radioactive probe
A single strand probe of complementary radioactive RNA or DNA is added, and the strands are allowed to re-anneal
Probe
Probe
Unbound probe washed away; filter exposed to film
Colonies that have annealed with the probe will expose the film; a comparison with the original petri dish reveals which colony is of interest
X-ray film

Figure B.25 Selecting a desired clone from a DNA library

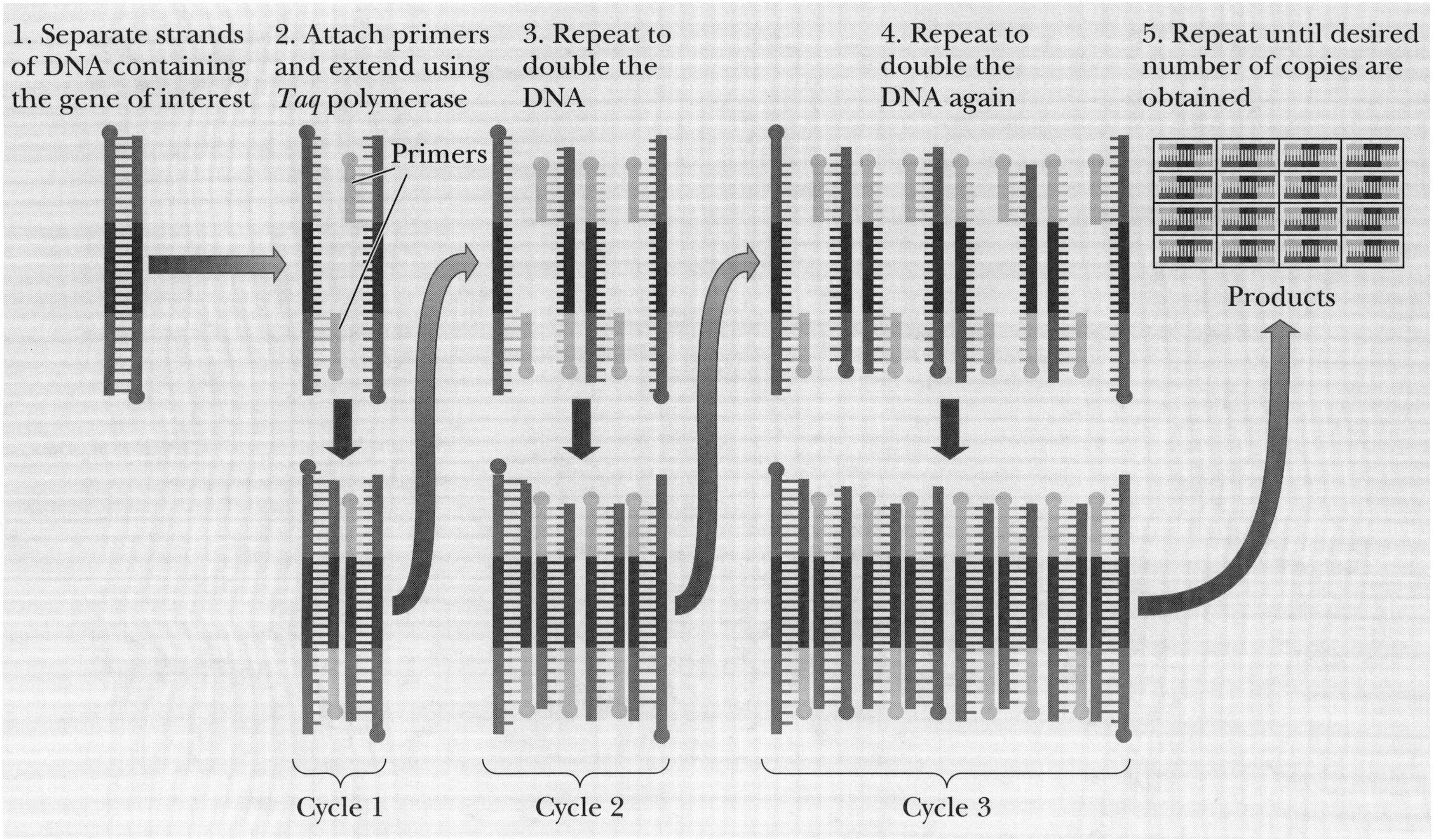
1. Separate strands of DNA containing the gene of interest
2. Attach primers and extend using *Taq* polymerase
Primers
3. Repeat to double the DNA
4. Repeat to double the DNA again
5. Repeat until desired number of copies are obtained
Products
Cycle 1
Cycle 2
Cycle 3

Figure B.26 The polymerase chain reaction

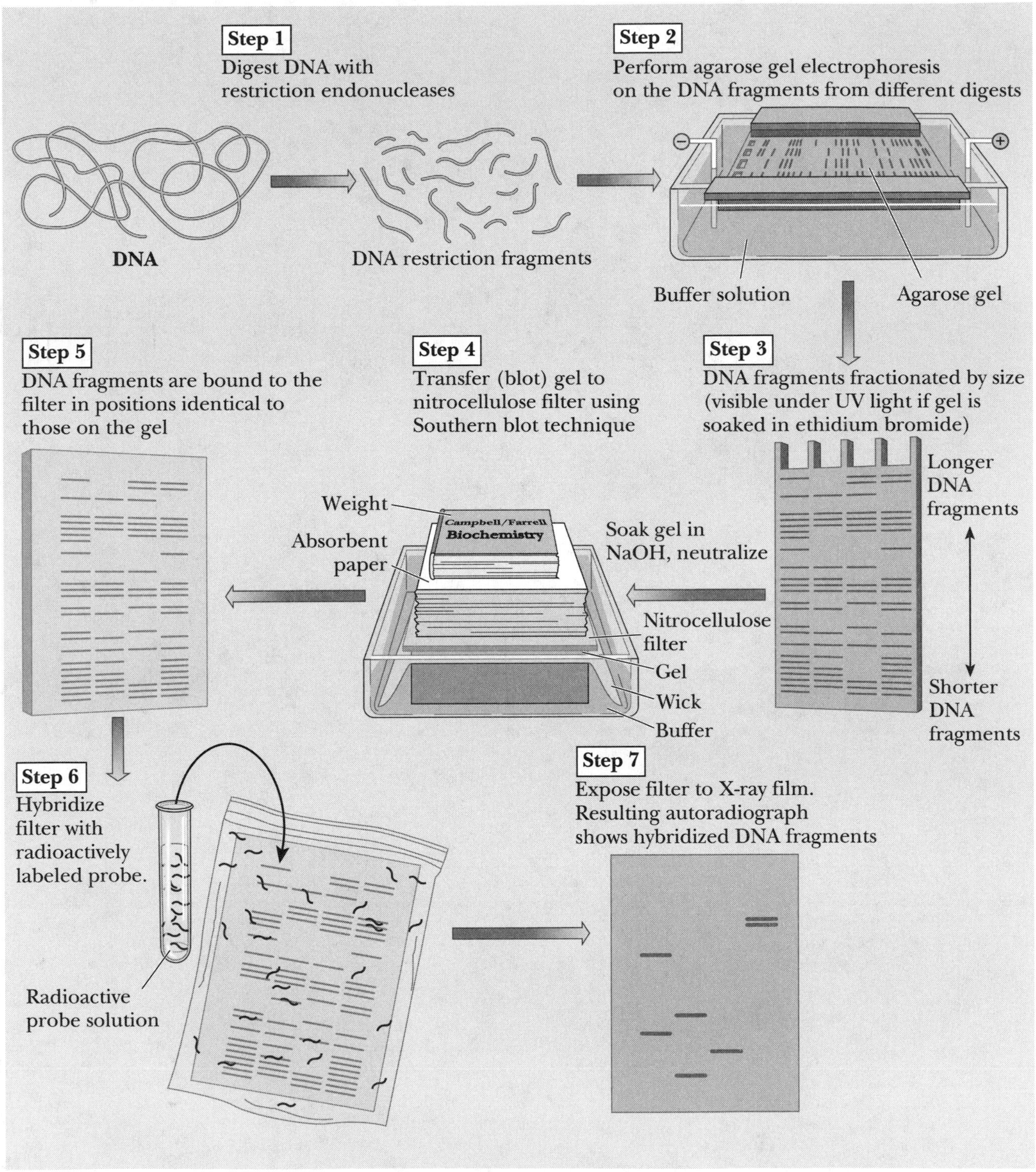
Step 1
Digest DNA with restriction endonucleases
DNA
DNA restriction fragments
Step 2
Perform agarose gel electrophoresis on the DNA fragments from different digests
Buffer solution
Agarose gel
Step 3
DNA fragments fractionated by size (visible under UV light if gel is soaked in ethidium bromide)
Longer DNA fragments
Shorter DNA fragments
Soak gel in NaOH, neutralize
Step 4
Transfer (blot) gel to nitrocellulose filter using Southern blot technique
Weight
Campbell/Farrell Biochemistry
Absorbent paper
Nitrocellulose filter
Gel
Wick
Buffer
Step 5
DNA fragments are bound to the filter in positions identical to those on the gel
Step 6
Hybridize filter with radioactively labeled probe.
Radioactive probe solution
Step 7
Expose filter to X-ray film. Resulting autoradiograph shows hybridized DNA fragments

Figure B.27 The Southern blot

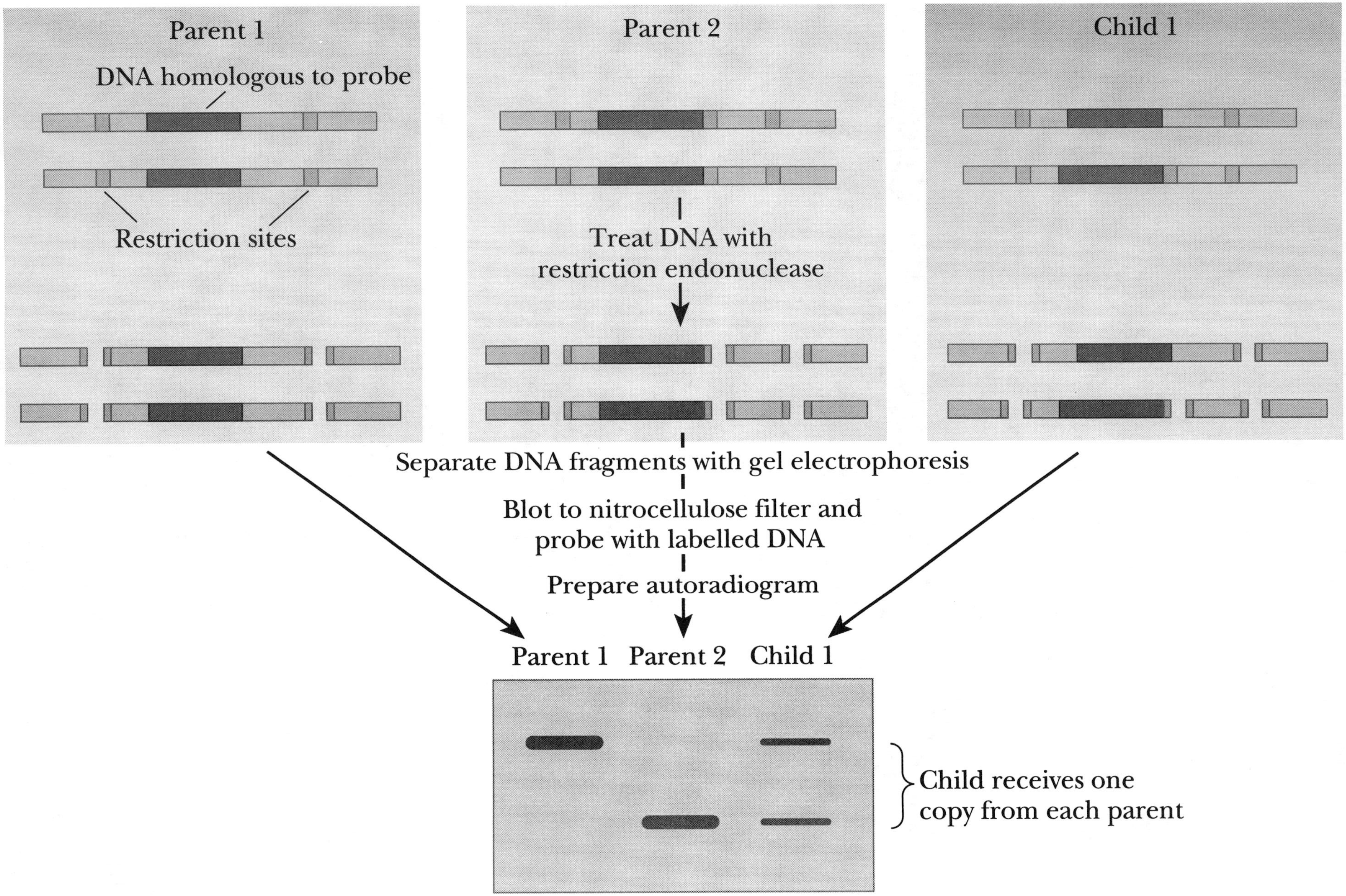

Figure B.29 Restriction fragment length polymorphisms

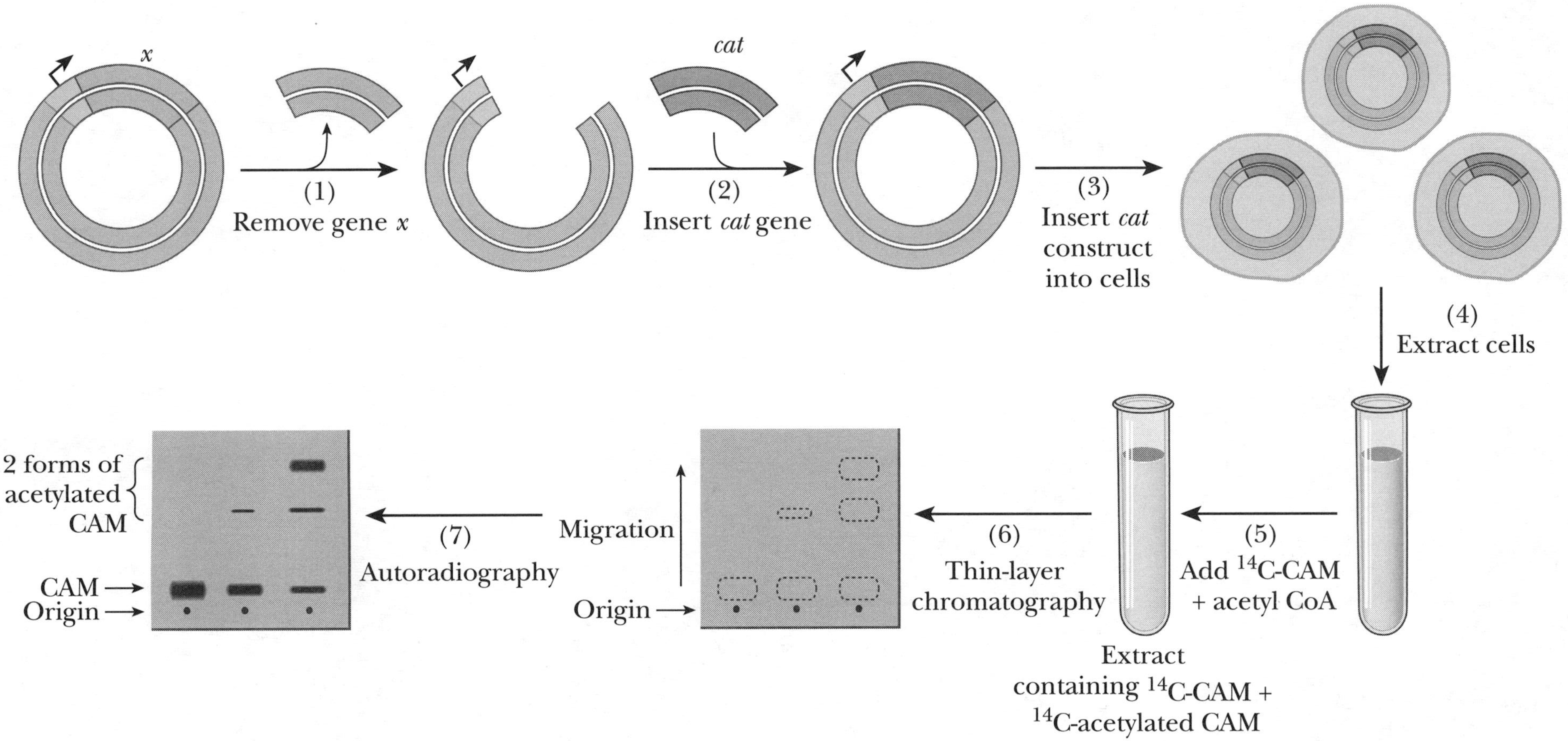

x
(1)
Remove gene x
cat
(2)
Insert cat gene
(3)
Insert cat construct into cells
(4)
Extract cells
(5)
Add ^{14}C-CAM + acetyl CoA
Extract containing ^{14}C-CAM + ^{14}C-acetylated CAM
(6)
Thin-layer chromatography
Migration
Origin
(7)
Autoradiography
2 forms of acetylated CAM
CAM
Origin

Figure B.34 Using a reporter gene

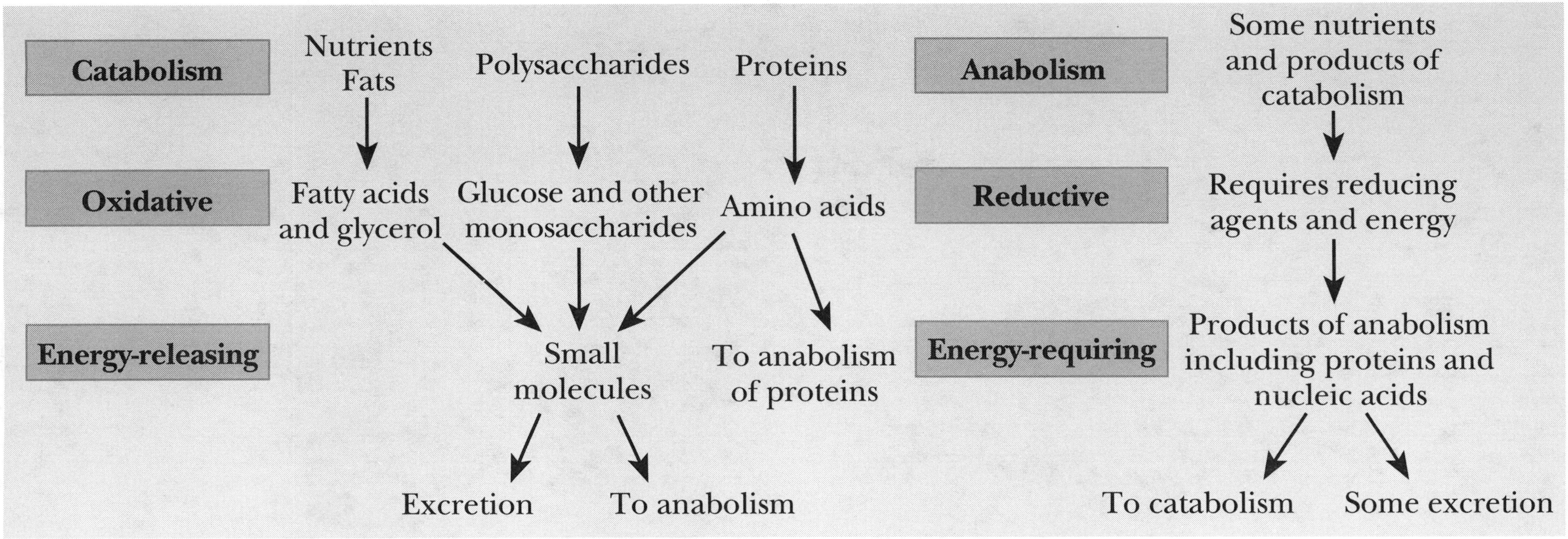

Figure 12.4 A comparison of catabolism and anabolism

(a) NADH
Nicotinamide ring
For NADH,
R = —H
For NADPH,
(b) NAD+
Nicotinamide ring
For NAD+,
R = —H
For NADP+,

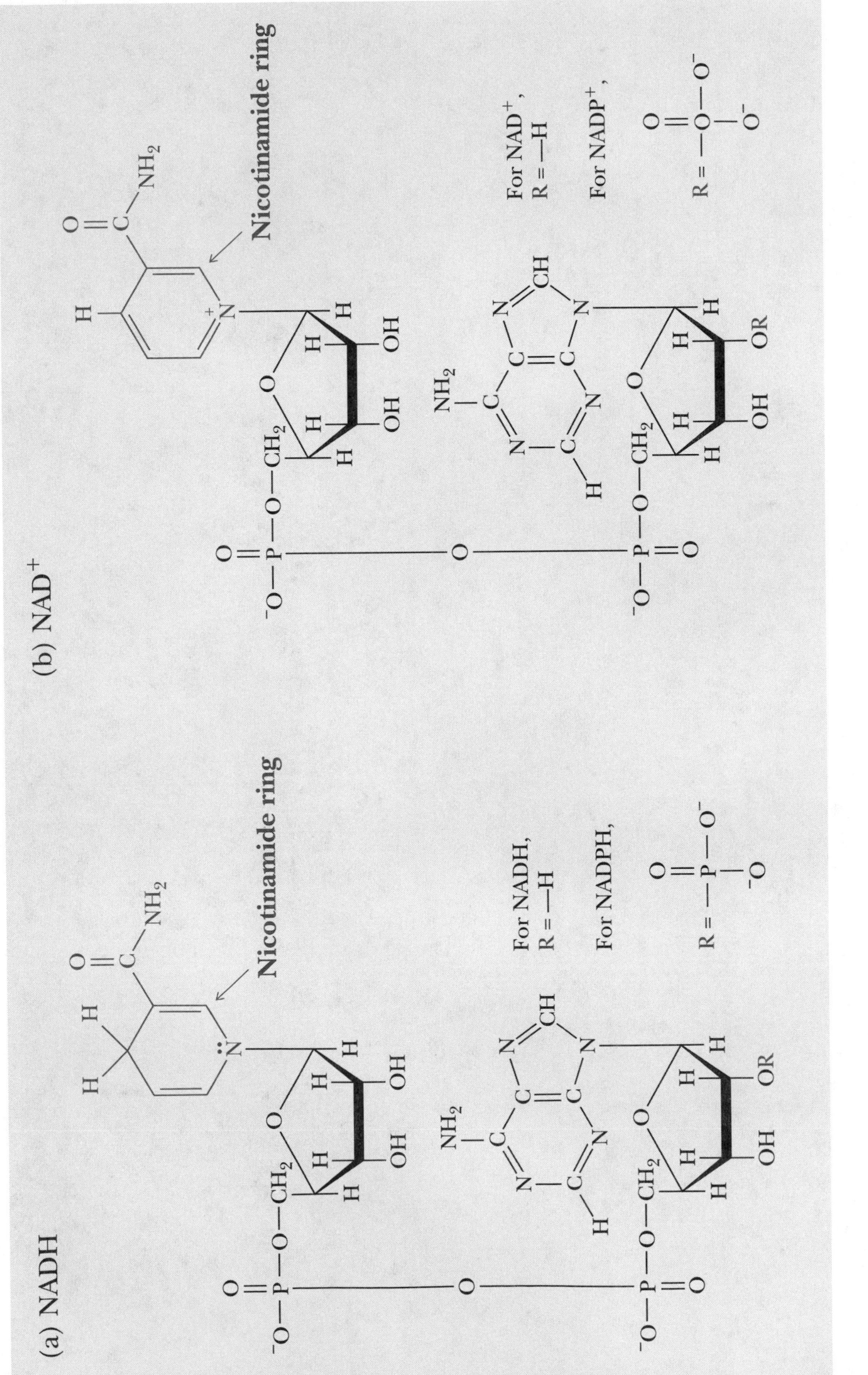

Figure 12.6 NADH and NAD$^+$

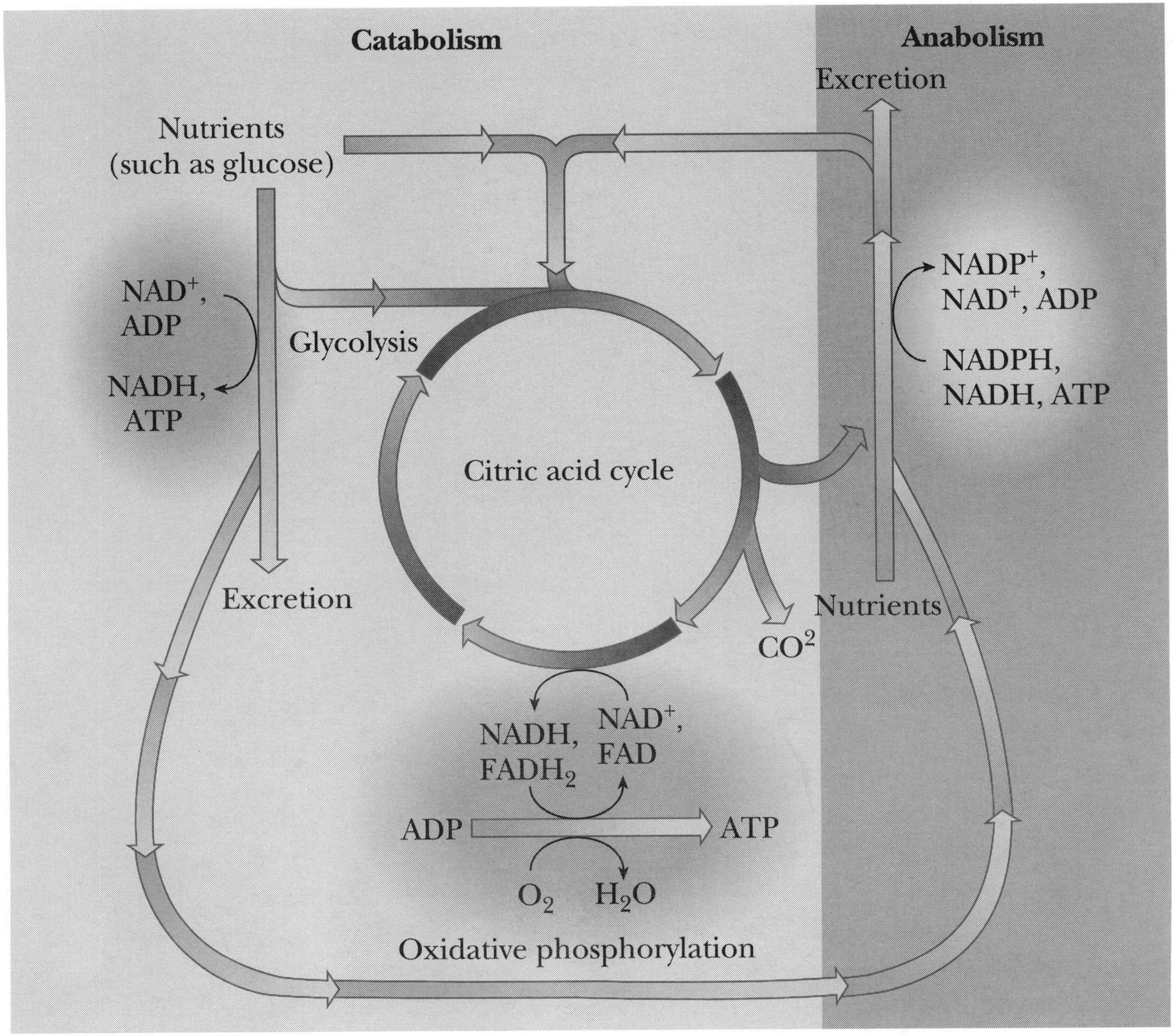

Catabolism
Anabolism
Excretion
Nutrients
(such as glucose)
NAD^+,
ADP
NADH,
ATP
Glycolysis
$NADP^+$,
NAD^+, ADP
NADPH,
NADH, ATP
Citric acid cycle
Excretion
Nutrients
CO^2
NADH,
$FADH_2$
NAD^+,
FAD
ADP
ATP
O_2
H_2O
Oxidative phosphorylation

Figure 12.14 Anabolism and catabolism

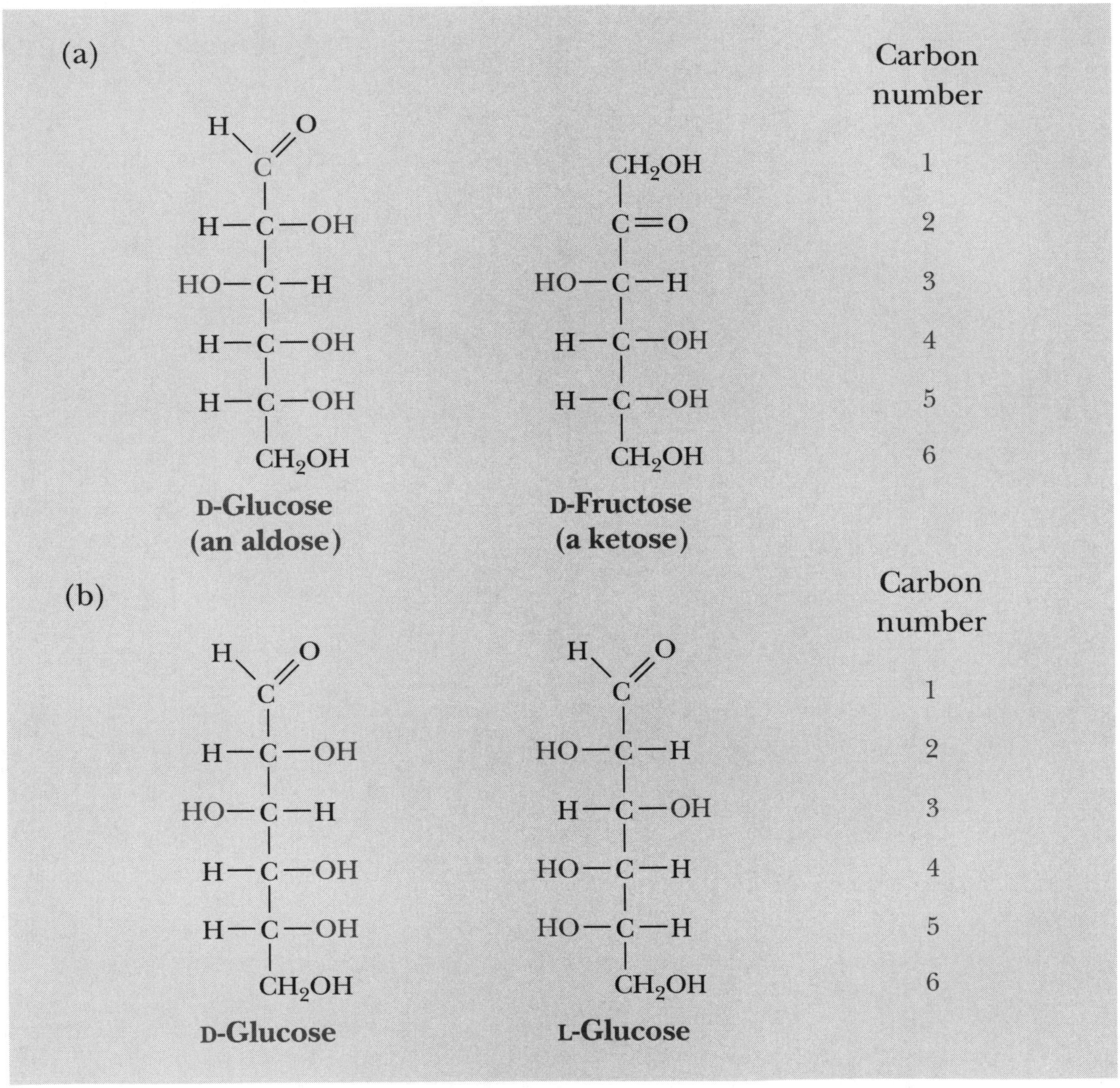
(a)
Carbon number
1
2
3
4
5
6
D-Glucose (an aldose)
D-Fructose (a ketose)
(b)
Carbon number
1
2
3
4
5
6
D-Glucose
L-Glucose

Figure 13.2 An aldose and a ketose

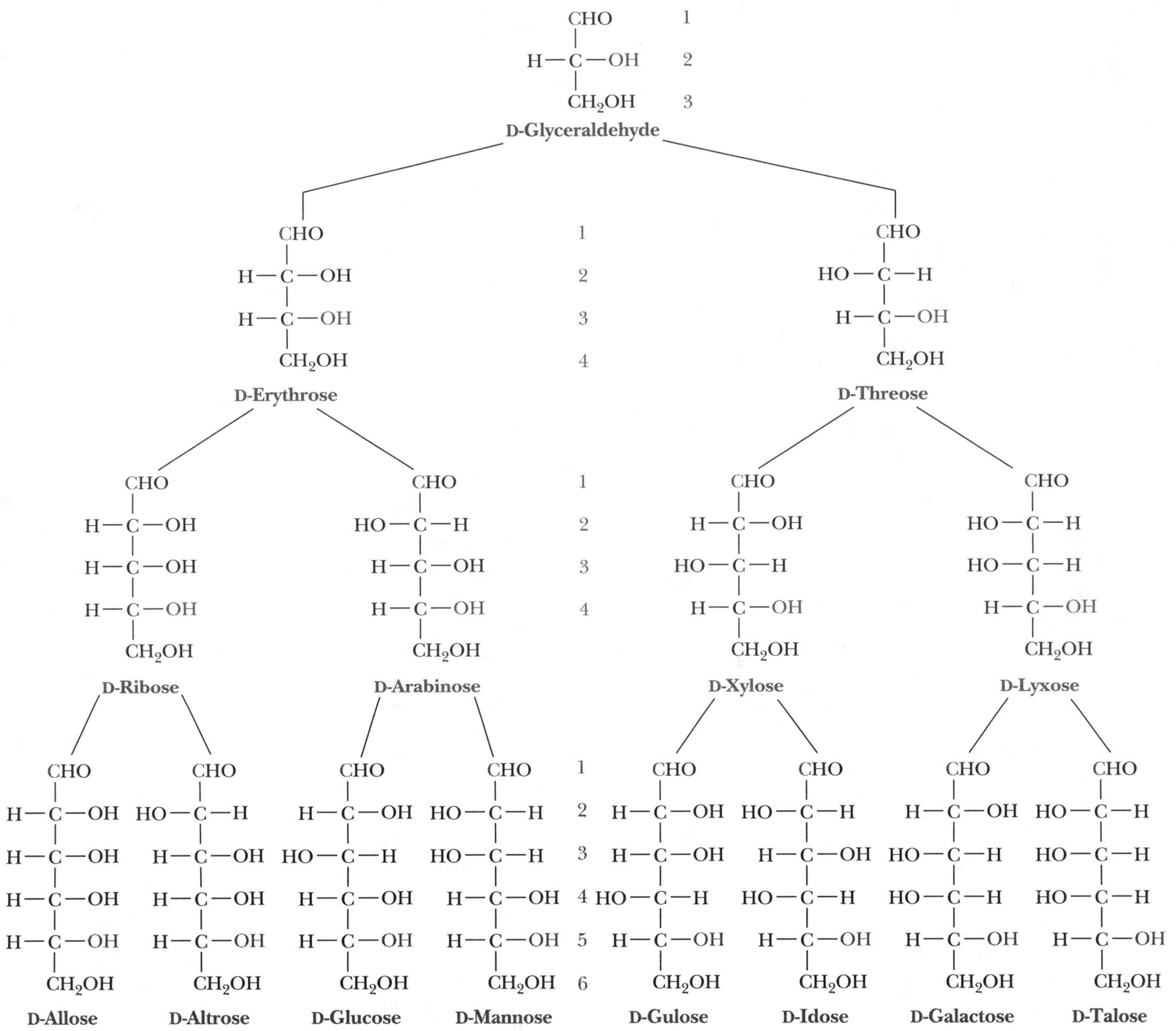

Figure 13.4 Stereochemical relationships among monosaccharides

Fischer	Complete Haworth	Abbreviated Haworth

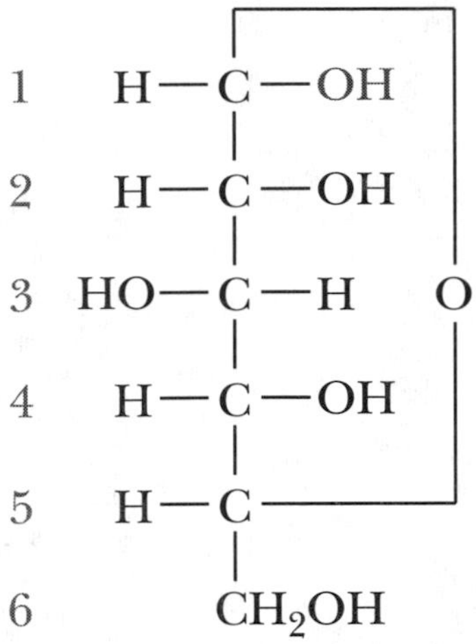
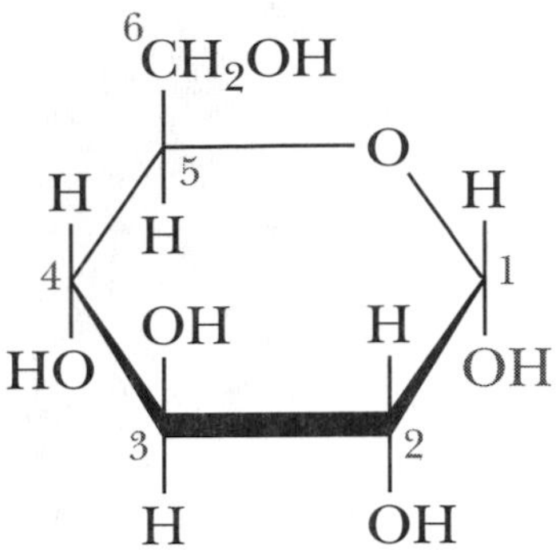
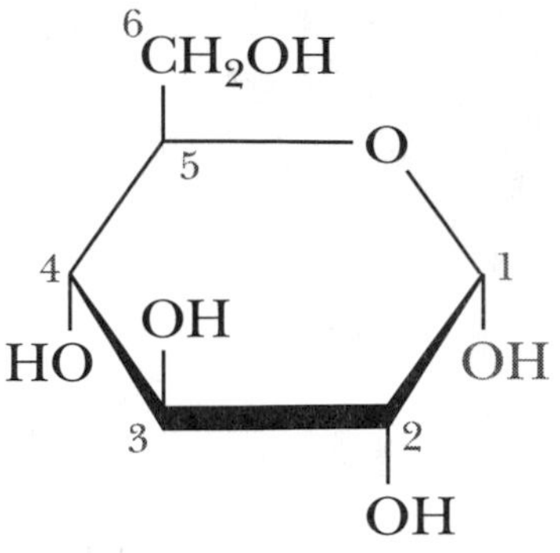

α-D-Glucose
(α-D-Glucopyranose)

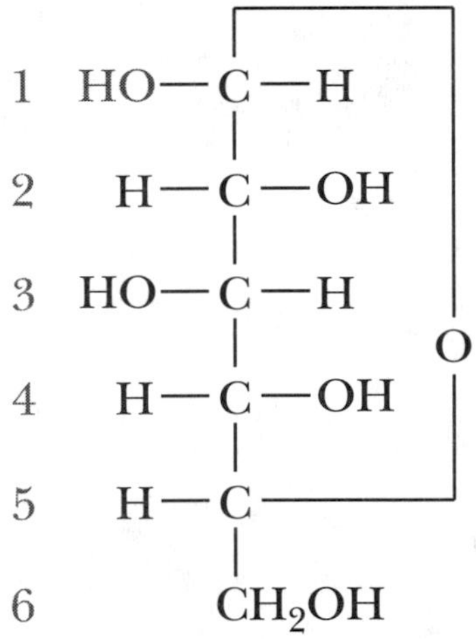
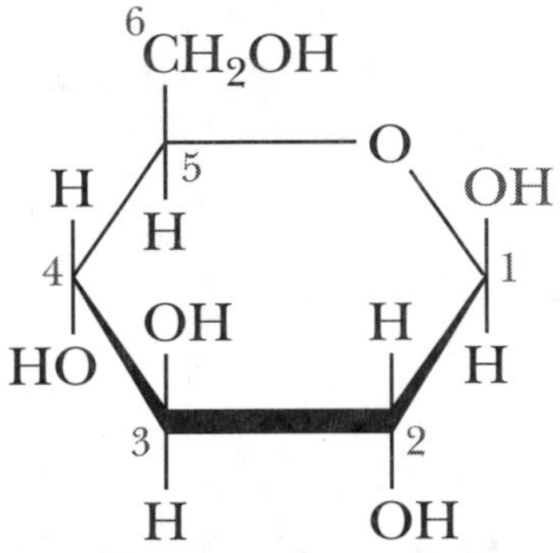
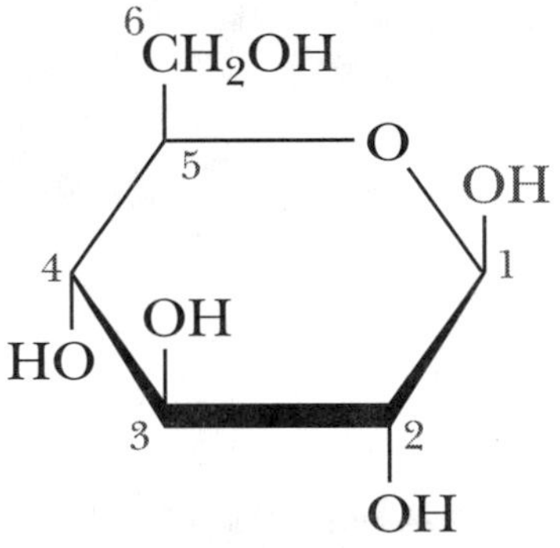

β-D-Glucose
(β-D-Glucopyranose)

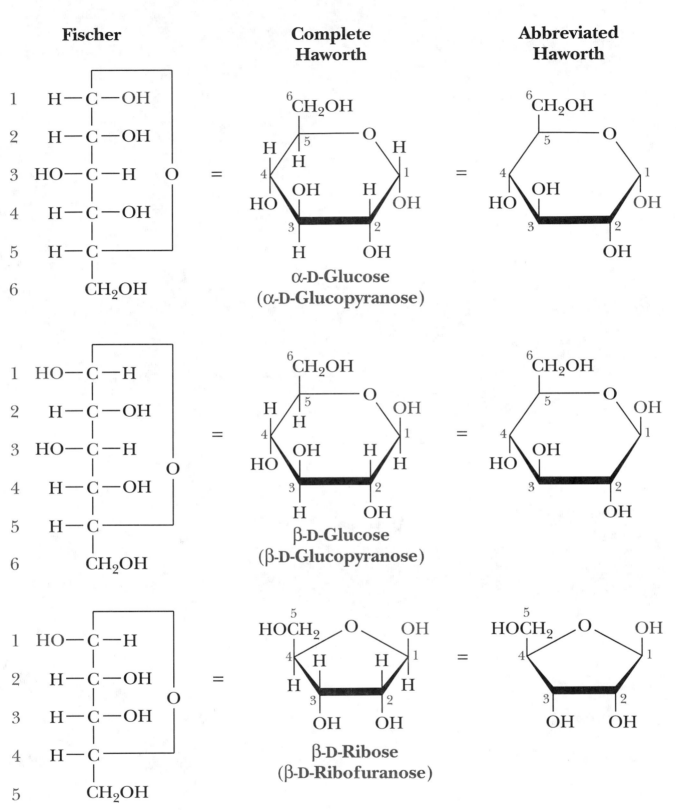

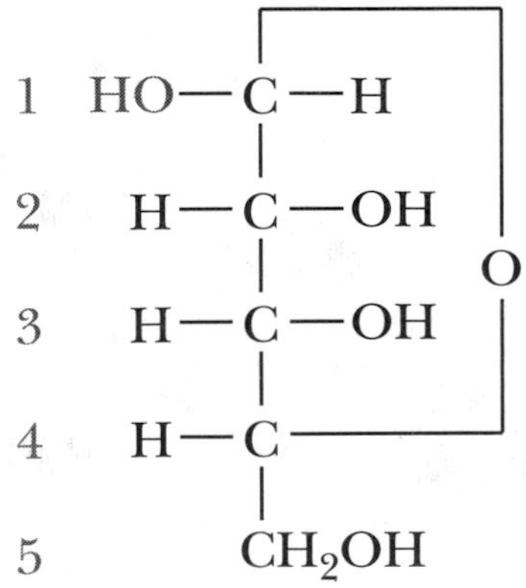
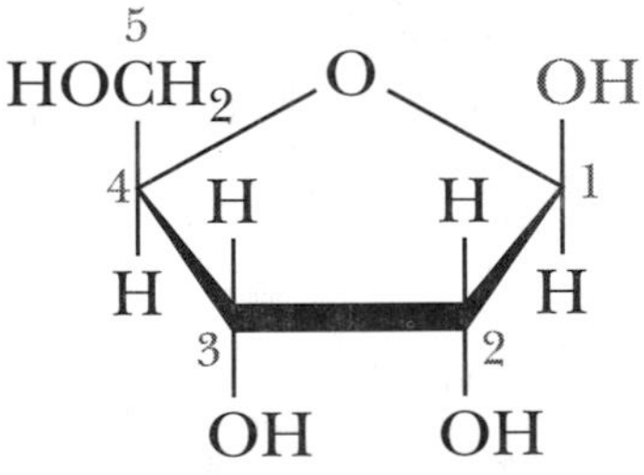
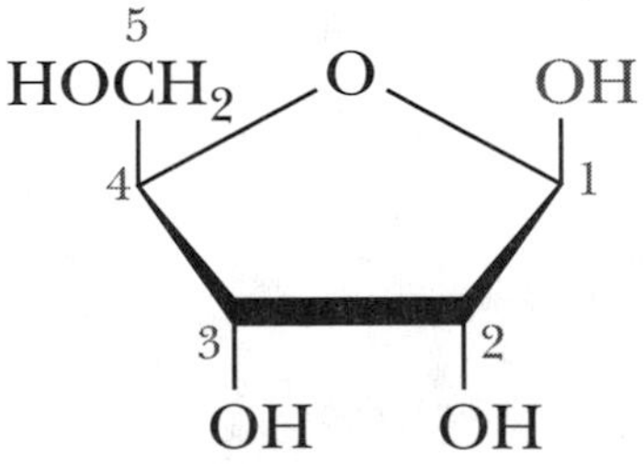

β-D-Ribose
(β-D-Ribofuranose)

Figure 13.8 A comparison of the Fischer, complete Haworth, and abbreviated Haworth representations of α- and β-D-glucose

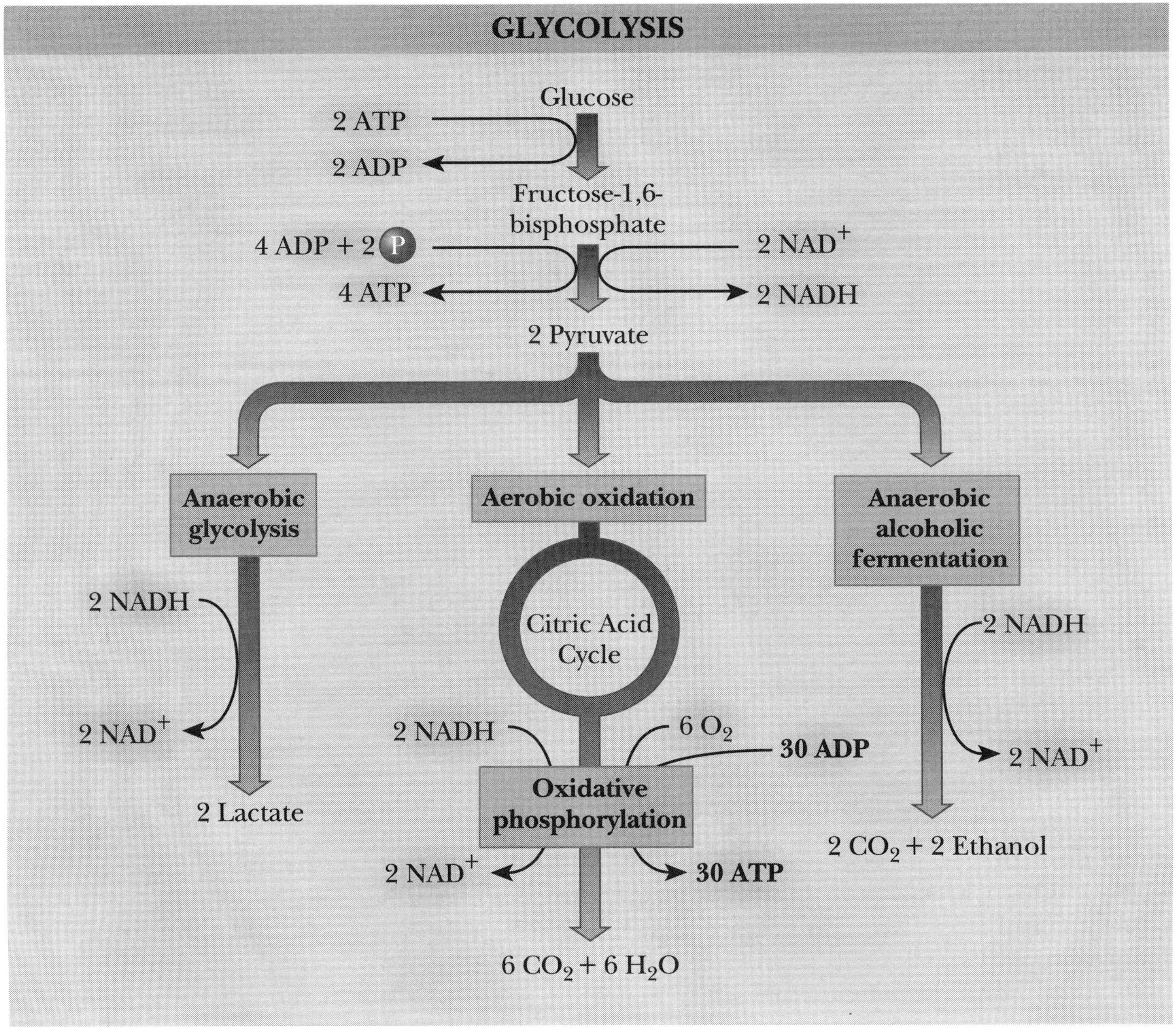

***Figure 14.1* One molecule of glucose is converted to two molecules of pyruvate**

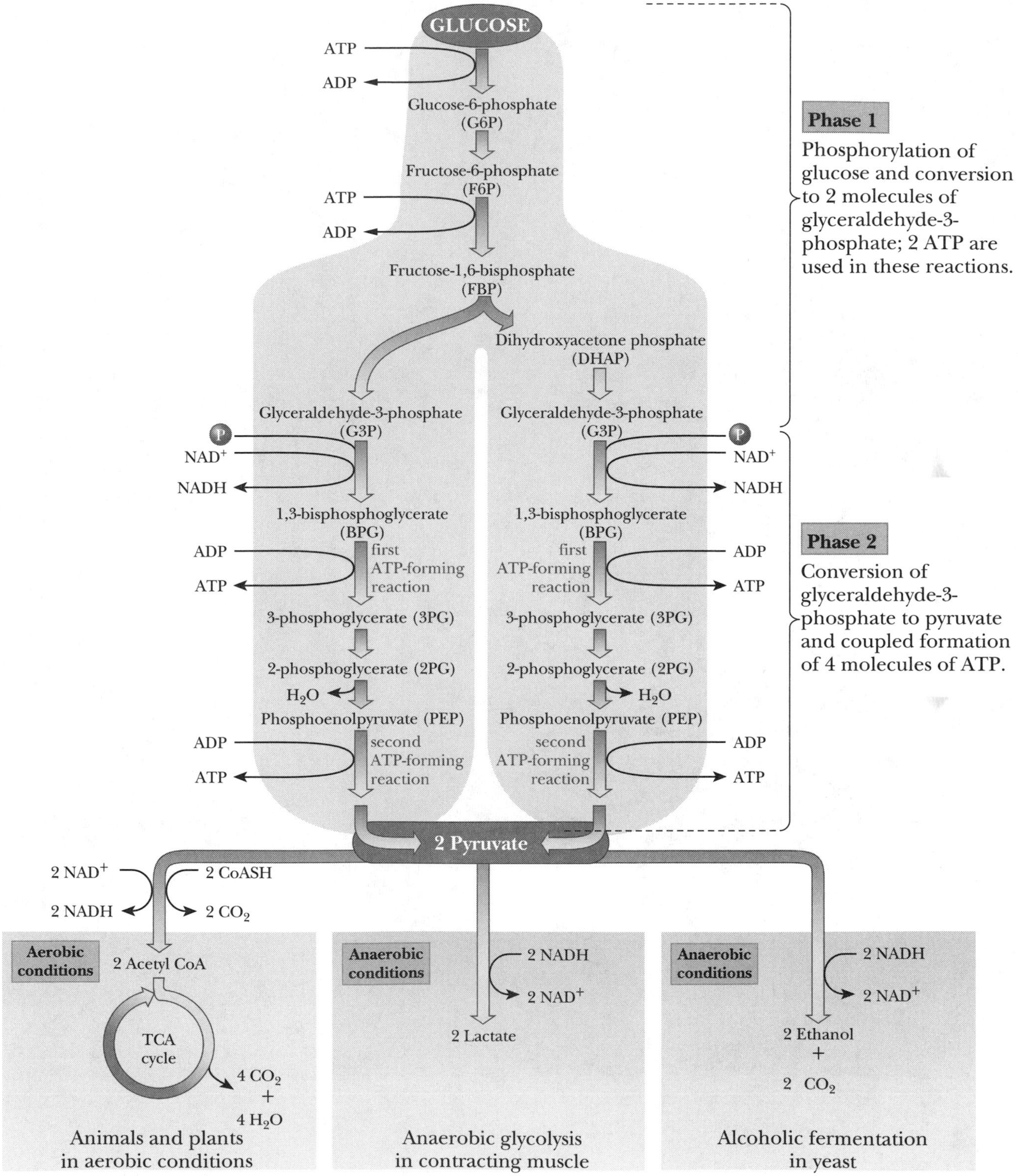
GLUCOSE
ATP
ADP
Glucose-6-phosphate (G6P)
Fructose-6-phosphate (F6P)
ATP
ADP
Fructose-1,6-bisphosphate (FBP)
Dihydroxyacetone phosphate (DHAP)
Glyceraldehyde-3-phosphate (G3P)
Glyceraldehyde-3-phosphate (G3P)
P
NAD+
NADH
1,3-bisphosphoglycerate (BPG)
ADP
ATP
first ATP-forming reaction
3-phosphoglycerate (3PG)
2-phosphoglycerate (2PG)
H2O
Phosphoenolpyruvate (PEP)
second ATP-forming reaction
2 Pyruvate
Phase 1
Phosphorylation of glucose and conversion to 2 molecules of glyceraldehyde-3-phosphate; 2 ATP are used in these reactions.
Phase 2
Conversion of glyceraldehyde-3-phosphate to pyruvate and coupled formation of 4 molecules of ATP.
2 NAD+
2 NADH
2 CoASH
2 CO2
Aerobic conditions
2 Acetyl CoA
TCA cycle
4 CO2
+
4 H2O
Animals and plants in aerobic conditions
Anaerobic conditions
2 NADH
2 NAD+
2 Lactate
Anaerobic glycolysis in contracting muscle
Anaerobic conditions
2 NADH
2 NAD+
2 Ethanol
+
2 CO2
Alcoholic fermentation in yeast

Figure 14.2 The pathway of glycolysis

In the first five steps of glycolysis, one six-carbon molecule of glucose is split into two 3-carbon compounds.

2 molecules of ATP are required for these reactions.

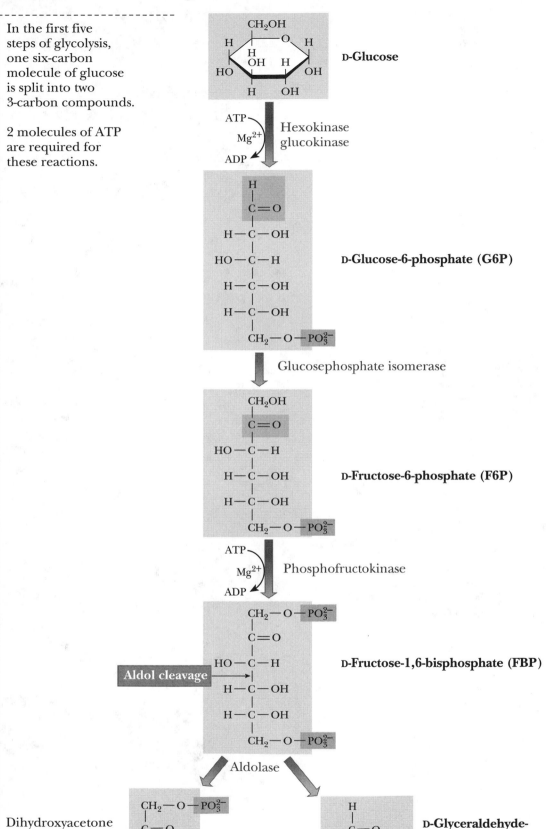

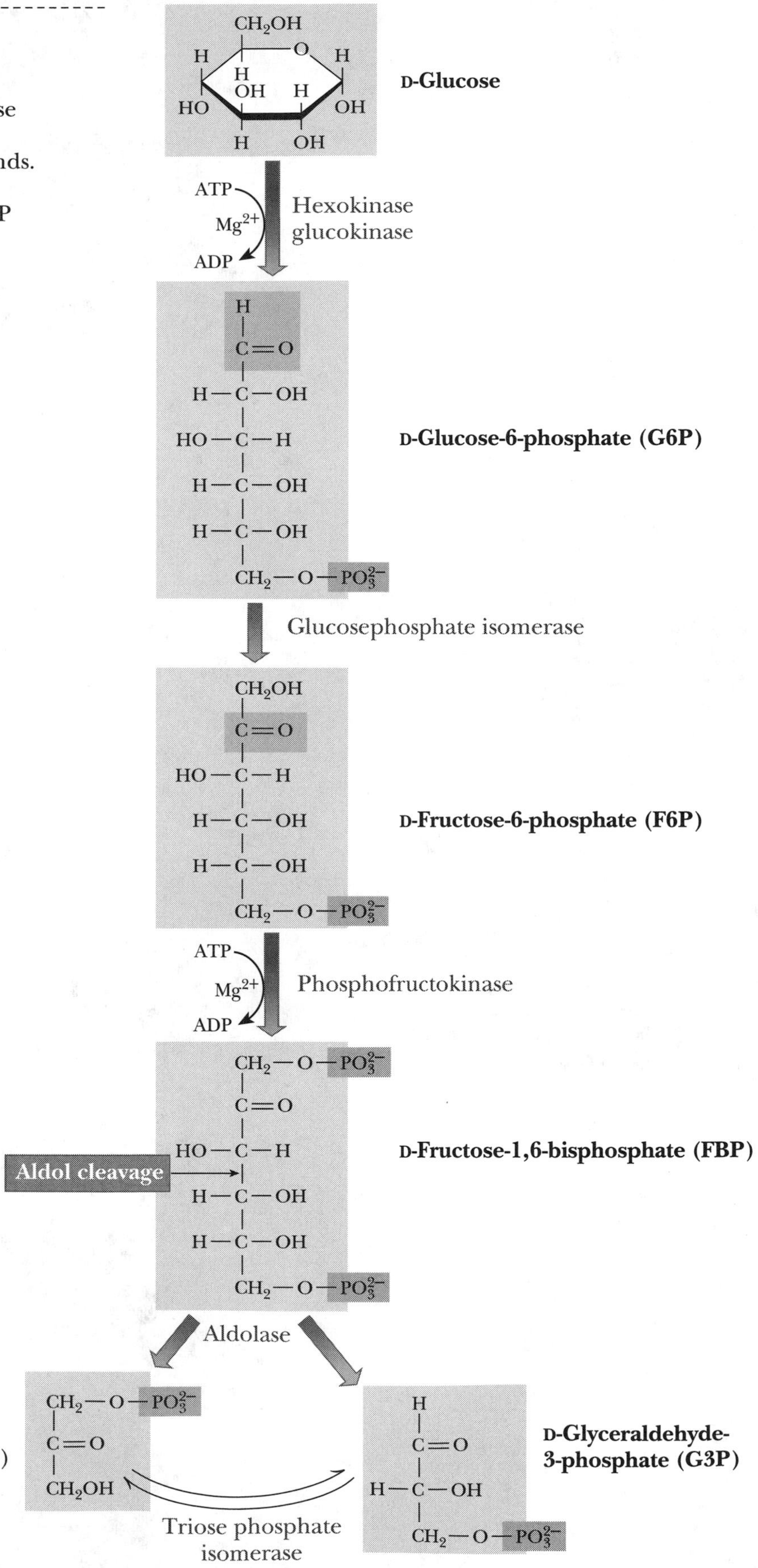

Figure 14.3 The first phase of glycolysis

In the second phase of glycolysis, glyceraldehyde-3-phosphate is converted to pyruvate.

These reactions yield 4 molecules of ATP, 2 for each molecule of pyruvate produced.

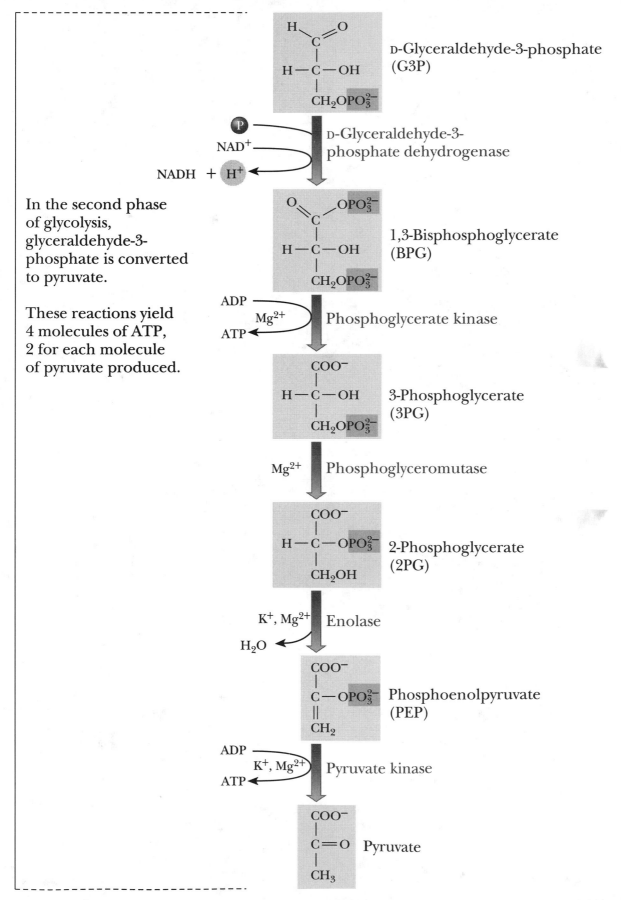

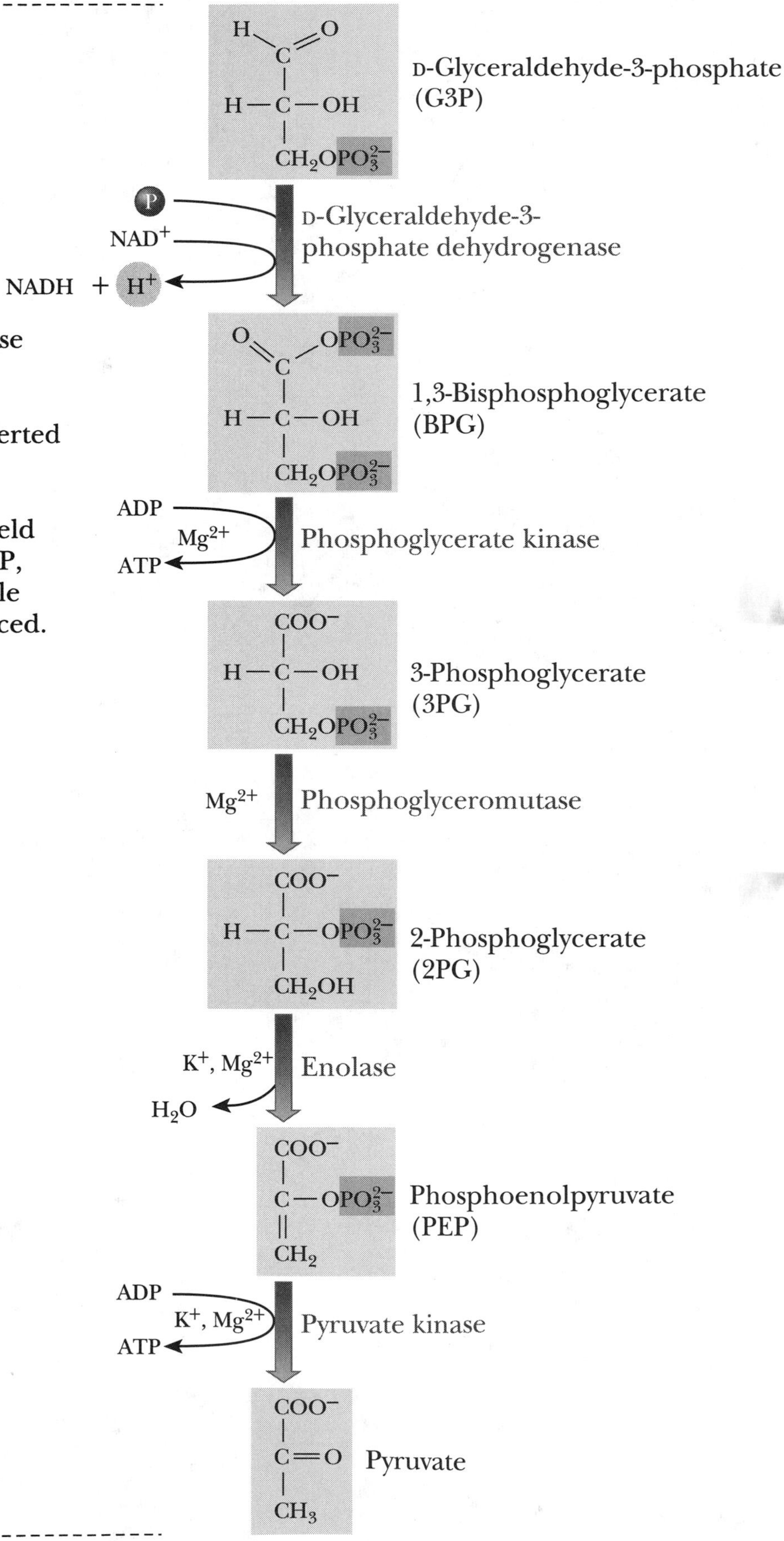

Figure 14.7 The second phase of glycolysis

TABLE 14.1 The Reactions of Glycolysis and Their Standard Free Energy Changes

Step	Reaction	Enzyme	$\Delta G^{\circ\prime}$*		ΔG**
			kJ mol^{-1}	*kcal mol*$^{-1}$	*kJ/mol*
1	Glucose + ATP → Glucose-6-phosphate + ADP	Hexokinase/ Glucokinase	−16.7	−4.0	−33.9
2	Glucose-6-phosphate → Fructose-6-phosphate	Glucose phosphate isomerase	+1.67	+0.4	−2.92
3	Fructose-6-phosphate + ATP → Fructose-1,6-*bis*phosphate + ADP	Phosphofructokinase	−14.2	−3.4	−18.8
4	Fructose-1,6-*bis*phosphate → Dihydroxyacetone phosphate + Glyceraldehyde-3-phosphate	Aldolase	+23.9	+5.7	−0.23
5	Dihydroxyacetone phosphate → Glyceraldehyde-3-phosphate	Triose phosphate isomerase	+7.56	+1.8	+2.41
6	2(Glyceraldehyde-3-phosphate + NAD^+ + P_i → 1,3-*bis*phosphoglycerate + NADH + H^+)	Glyceraldehyde-3-P dehydrogenase	2(+6.20)	2(+1.5)	2(−1.29)
7	2(1,3-*Bis*phosphoglycerate + ADP → 3-Phosphoglycerate + ATP)	Phosphoglycerate kinase	2(−18.8)	2(−4.5)	2.(+0.1)
8	2(3-Phosphoglycerate → 2-Phosphoglycerate)	Phosphoglyceromutase	2(+4.4)	2(+1.1)	2.(+0.83)
9	2(2-Phosphoglycerate → Phosphoenolpyruvate + H_2O)	Enolase	2(1.8)	2(+0.4)	2(+1.1)
10	2(Phosphoenolpyruvate + ADP → Pyruvate ATP)	Pyruvate kinase	2(−31.4)	2(−7.5)	2(−23.0)
Overall	Glucose + 2 ADP + 2 P_i + NAD^+ → 2 Pyruvate + 2 ATP + NADH + H^+	Lactate dehydrogenase	−73.3	−17.5	−98.0
	2(Pyruvate + NADH + H^+ → Lactate + NAD^+		2(−25.1)	2(−6.0)	2(−14.8)
	Glucose + 2 ADP + 2 P_i → 2 Lactate + 2 ATP		−123.5	−29.5	−127.6

*$\Delta G^{\circ\prime}$ values are assumed to be the same at 25°C and 37°C and are calculated for standard-state conditions (1 *M* concentration of reactants and products pH 7.0).

**ΔG values are calculated at 310 K (37°C) using steady-state concentrations of these metabolites found in erythrocytes.

Table 14.1 The reactions of glycolysis and their standard free energy changes

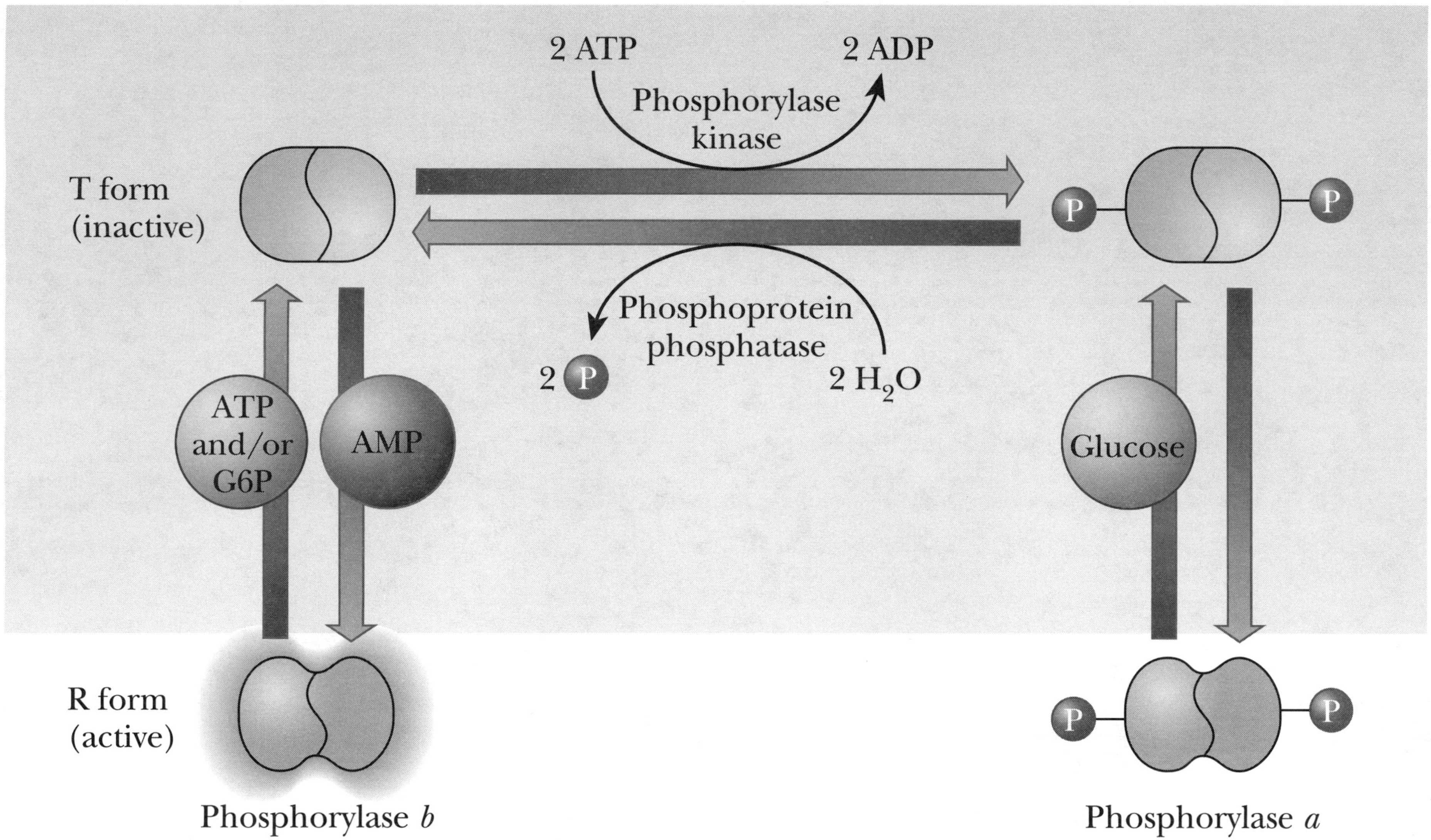

Figure 15.5 Glycogen phosphorylase activity is subject to allosteric control and covalent modification

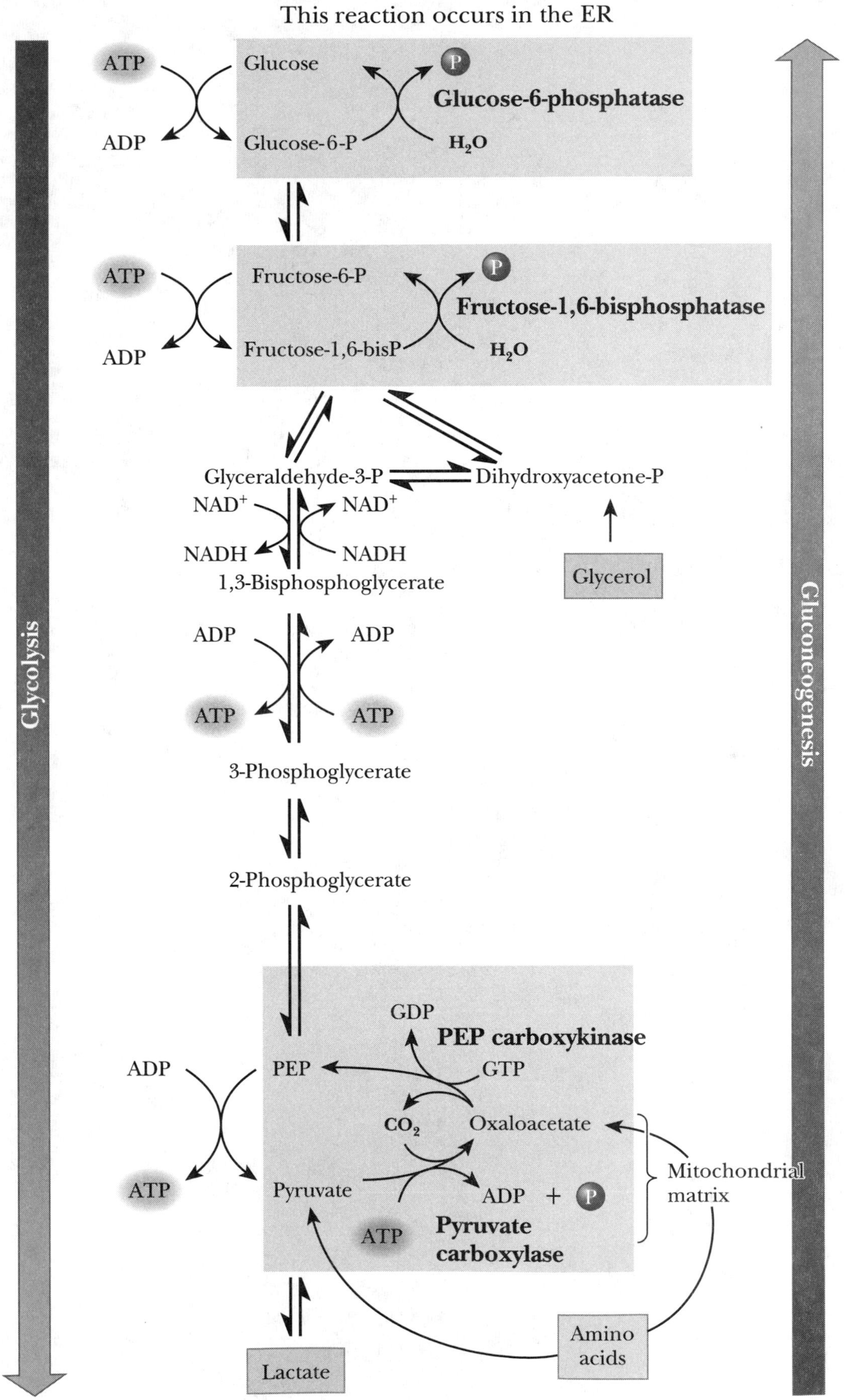

Figure 15.6 The pathways of gluconeogenesis and glycolysis

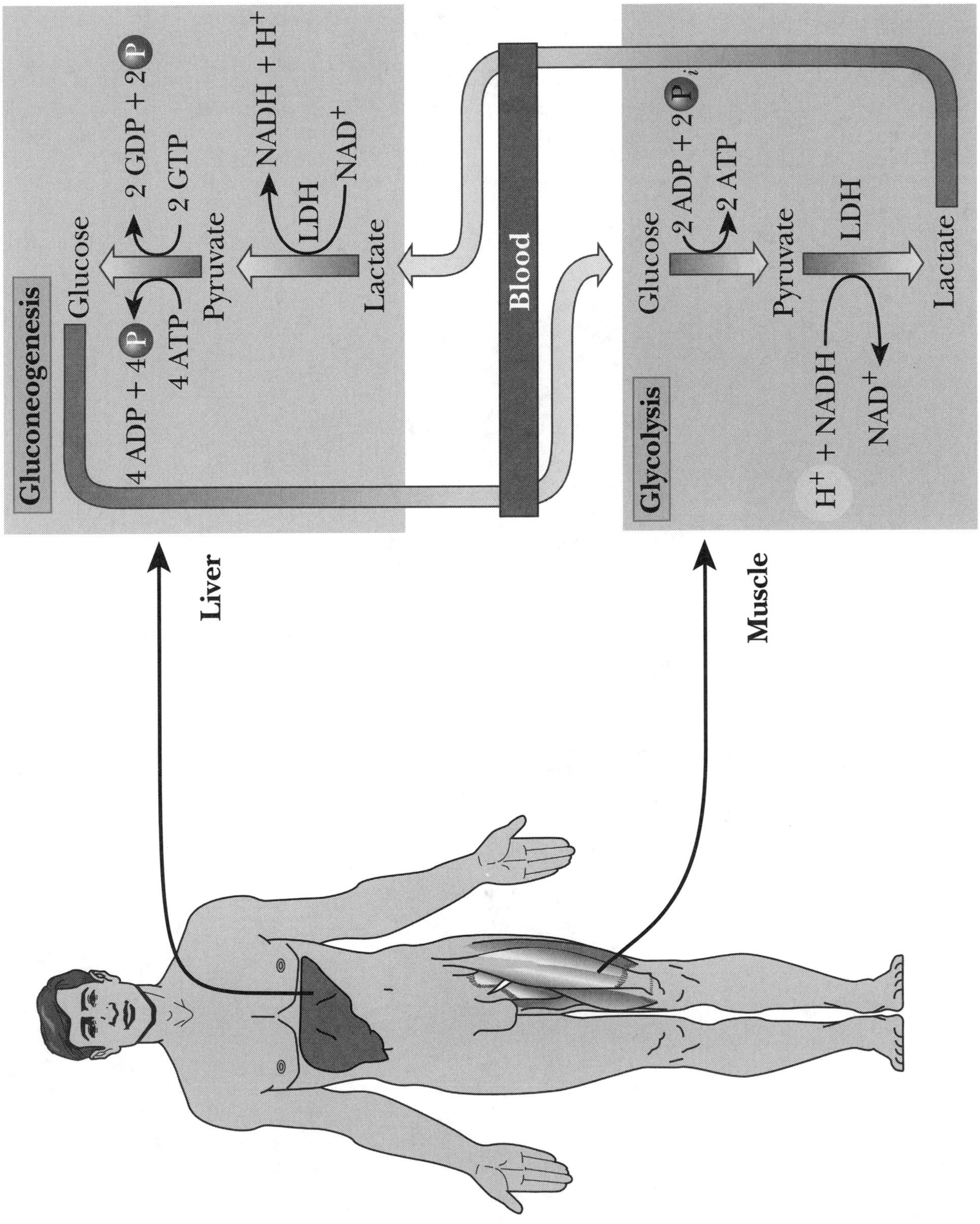
Liver
Muscle
Gluconeogenesis
Glucose
4 ADP + 4 P
4 ATP
2 GDP + 2 P
2 GTP
Pyruvate
NADH + H+
LDH
NAD+
Lactate
Blood
Glycolysis
Glucose
2 ADP + 2 Pi
2 ATP
Pyruvate
H+ + NADH
NAD+
LDH
Lactate

Figure 15.12 The Cori cycle

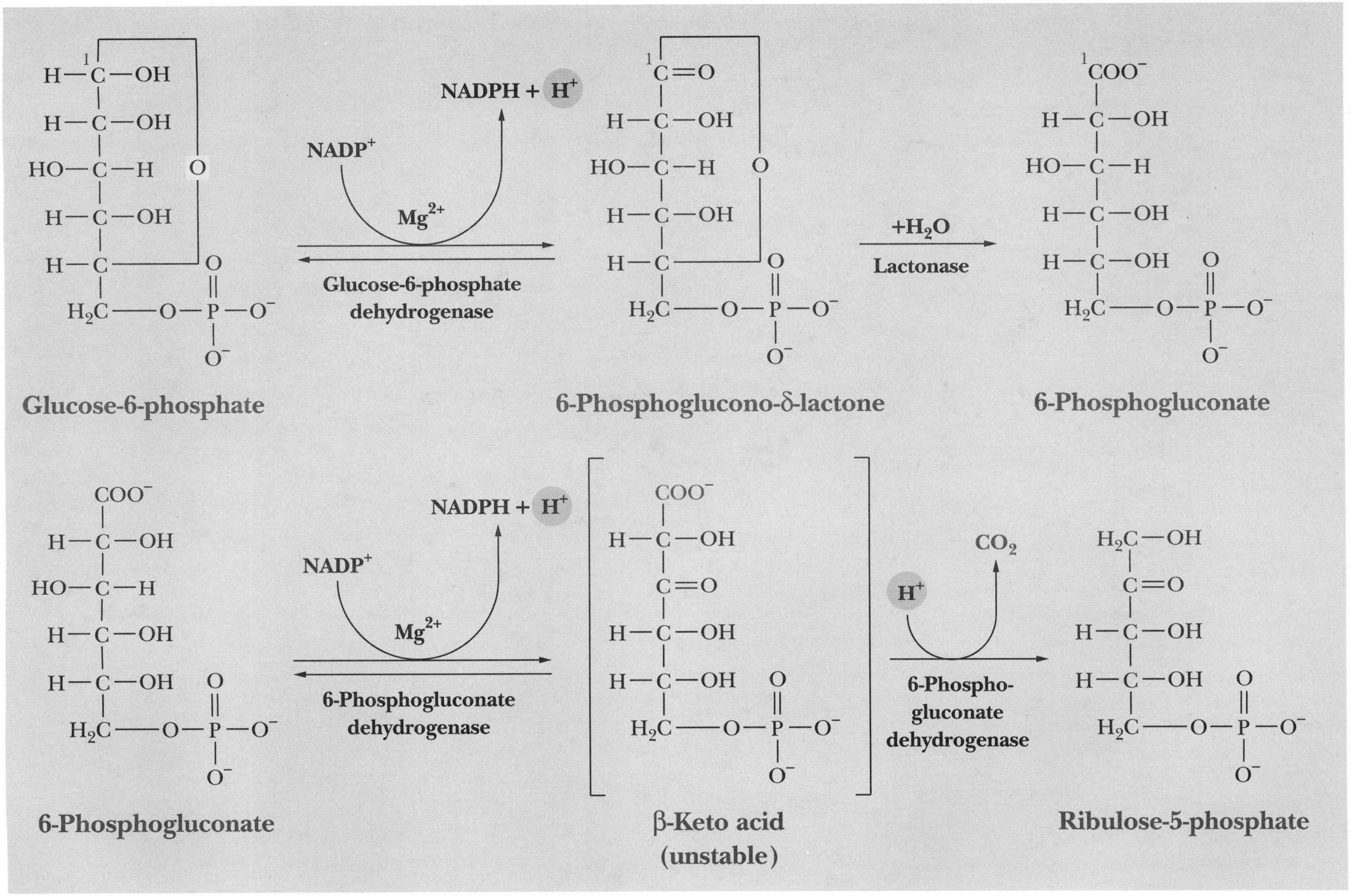

Figure 15.15 The oxidative reactions of the pentose phosphate pathway

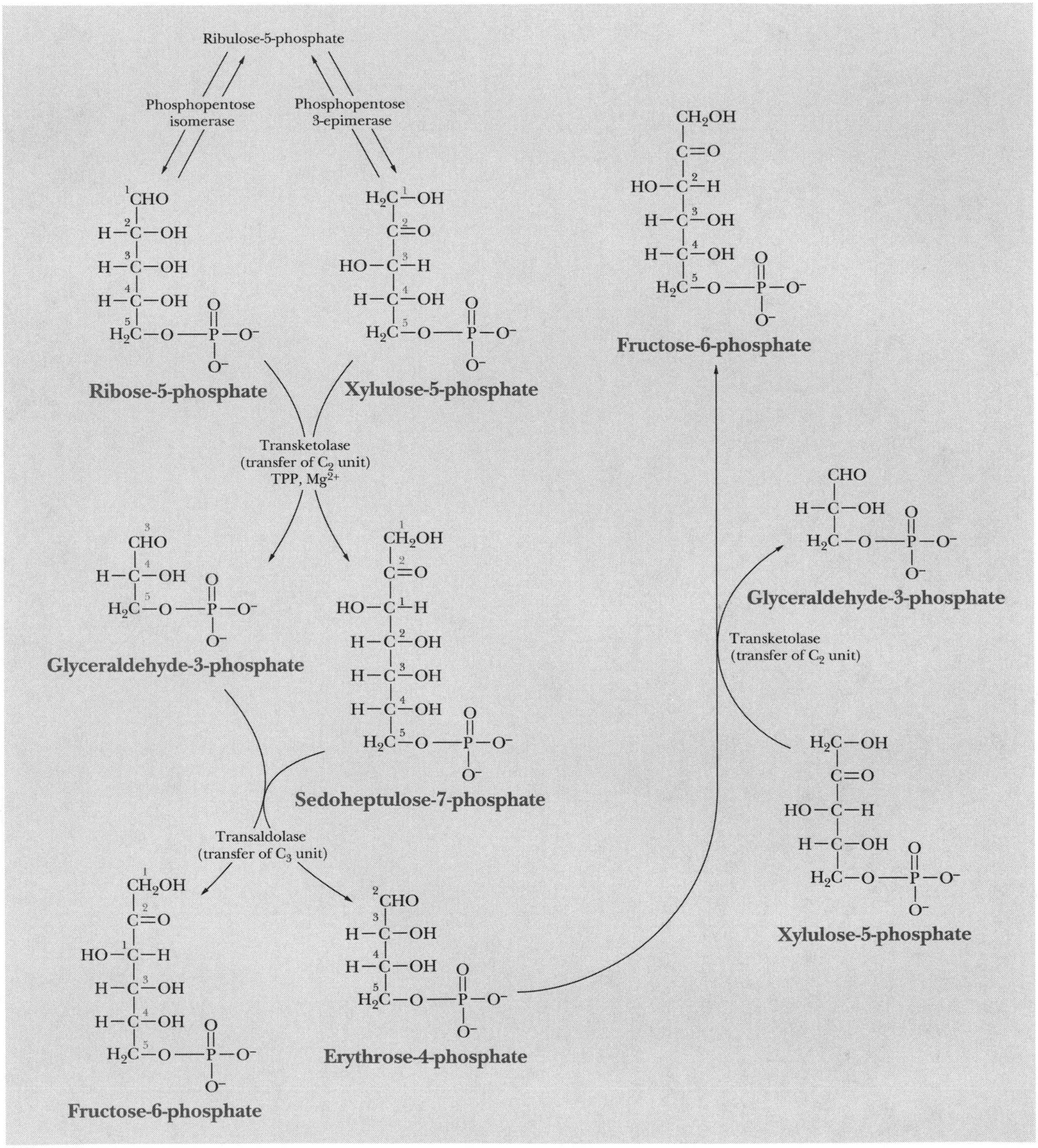

Figure 15.16 The nonoxidative reactions of the pentose phosphate pathway

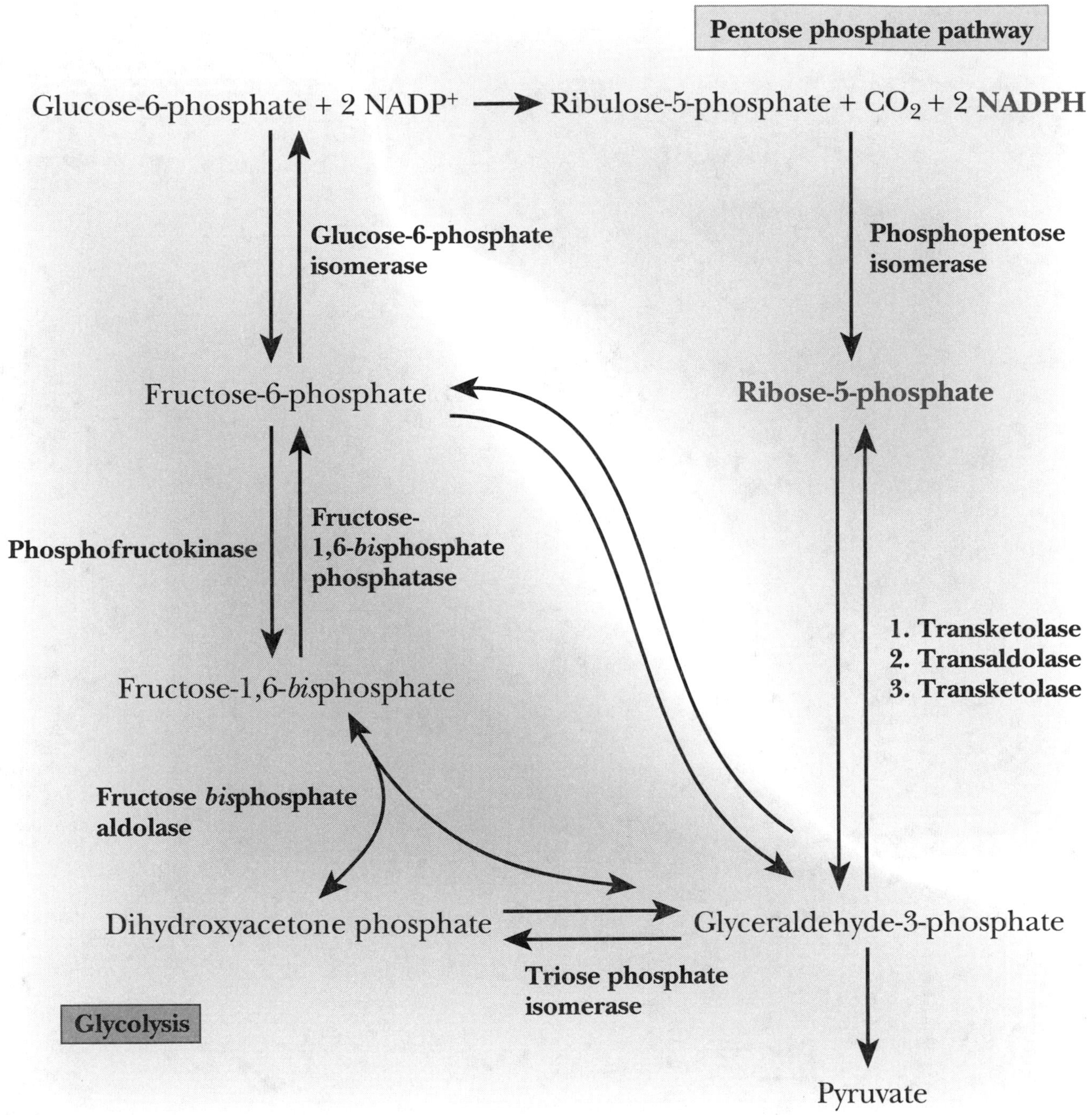

Figure 15.17 Relationship between the pentose phosphate pathway and glycolysis

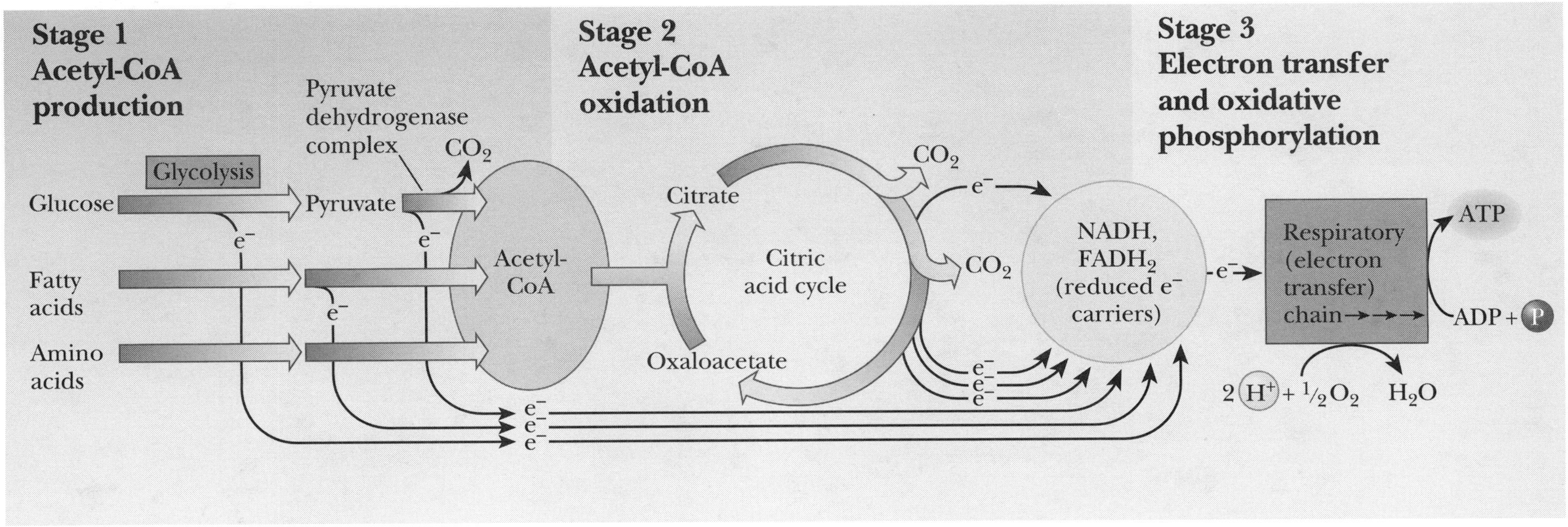
Stage 1
Acetyl-CoA
production
Stage 2
Acetyl-CoA
oxidation
Stage 3
Electron transfer
and oxidative
phosphorylation
Glycolysis
Glucose
Fatty
acids
Amino
acids
Pyruvate
dehydrogenase
complex
Pyruvate
CO_2
e^-
Acetyl-
CoA
Citrate
Citric
acid cycle
Oxaloacetate
CO_2
CO_2
NADH,
$FADH_2$
(reduced e^-
carriers)
Respiratory
(electron
transfer)
chain
ATP
ADP + P
$2 H^+ + 1/2 O_2$
H_2O

Figure 16.1 The citric acid cycle and catabolism

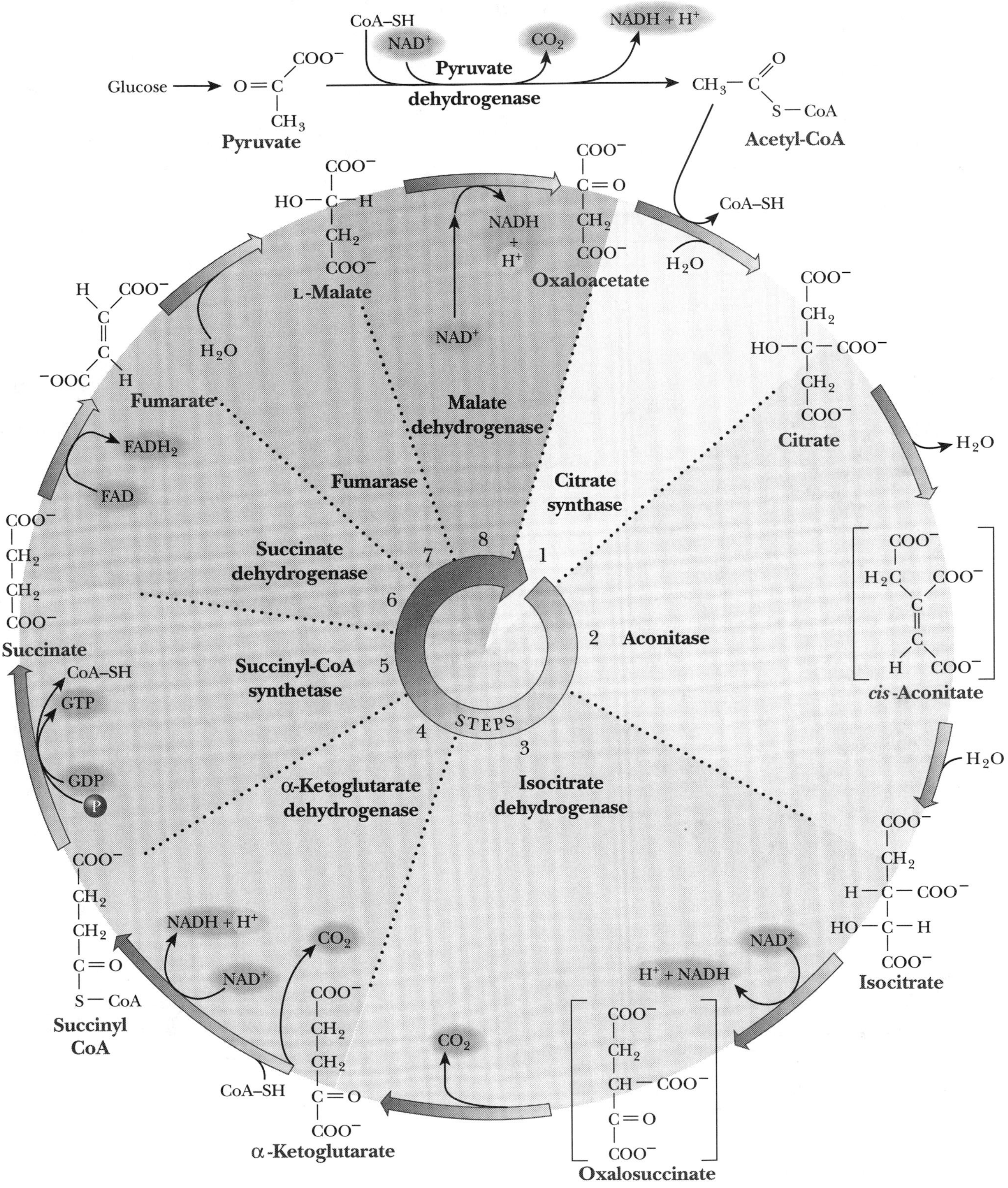

Glucose
Pyruvate
CoA–SH
NAD⁺
Pyruvate dehydrogenase
CO₂
NADH + H⁺
Acetyl-CoA
CoA–SH
H₂O
Oxaloacetate
NADH + H⁺
NAD⁺
L-Malate
H₂O
Fumarate
FADH₂
FAD
Succinate
CoA–SH
GTP
GDP
P
Succinyl CoA
NADH + H⁺
NAD⁺
CO₂
CoA–SH
α-Ketoglutarate
CO₂
Oxalosuccinate
NAD⁺
H⁺ + NADH
Isocitrate
H₂O
cis-Aconitate
H₂O
Citrate
Citrate synthase
Aconitase
Isocitrate dehydrogenase
α-Ketoglutarate dehydrogenase
Succinyl-CoA synthetase
Succinate dehydrogenase
Fumarase
Malate dehydrogenase
STEPS
1 2 3 4 5 6 7 8

Figure 16.3 An overview of the citric acid cycle

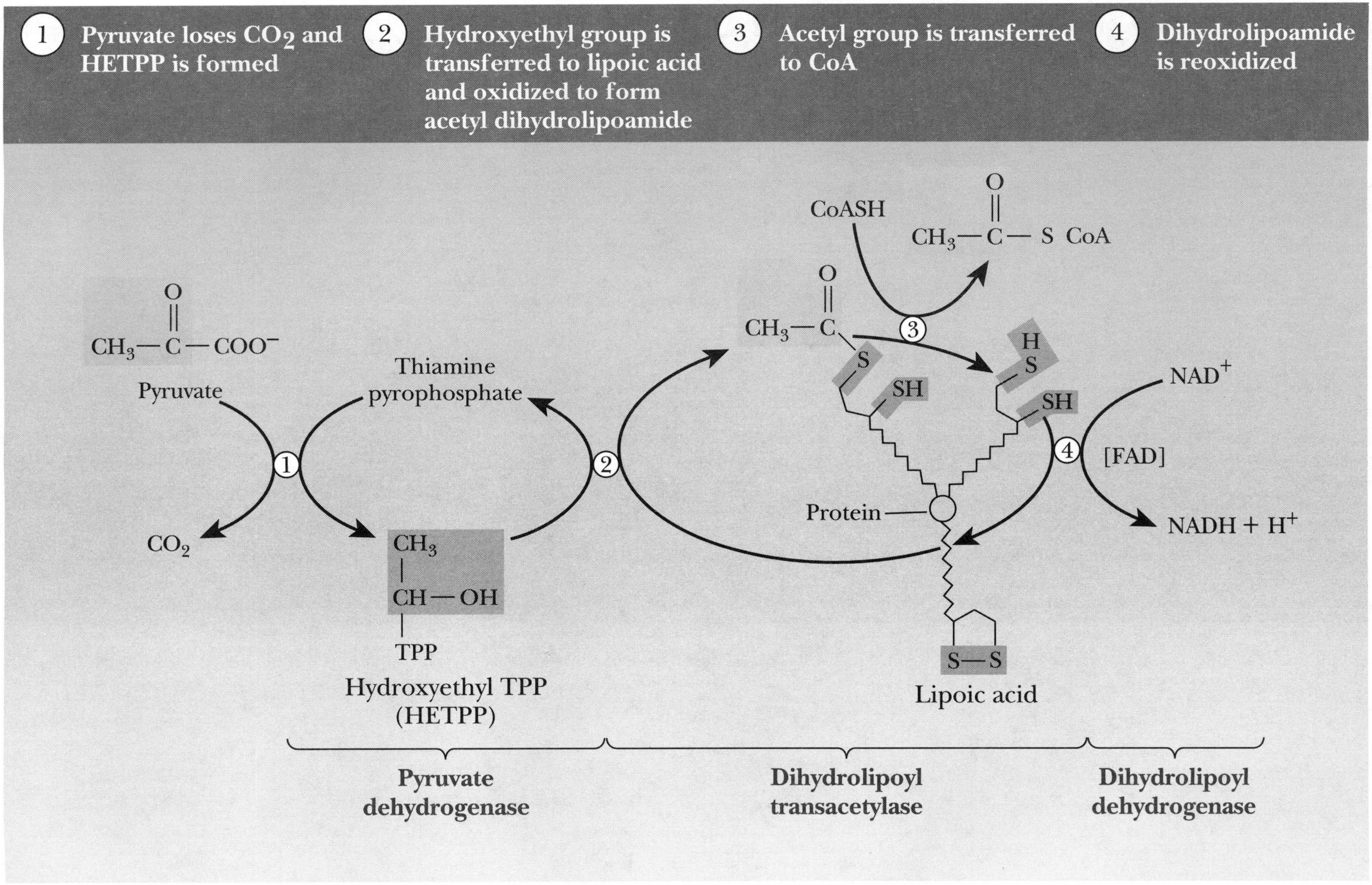

Figure 16.4 The mechanism of the pyruvate dehydrogenase reaction

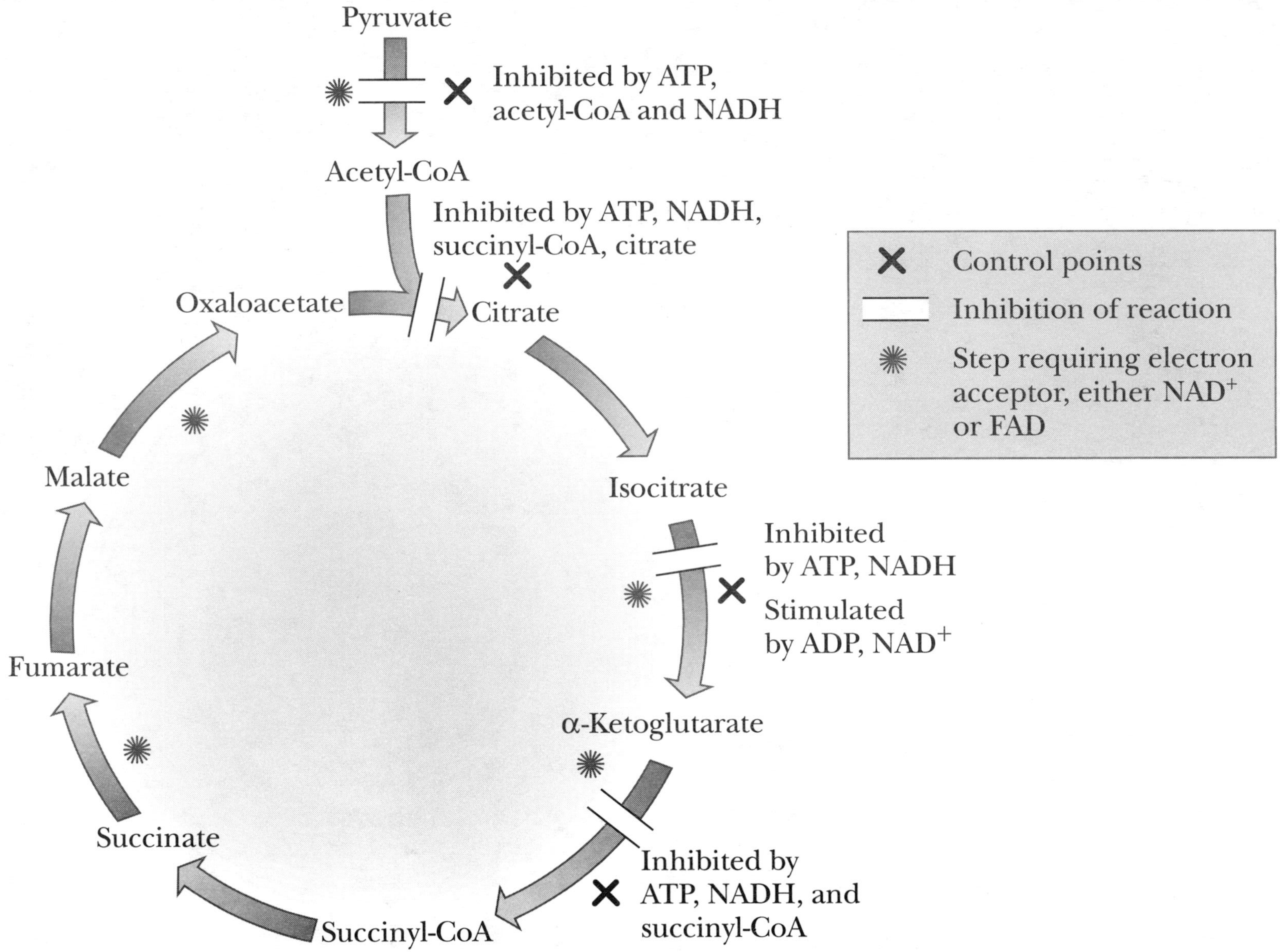

Figure 16.7 Control points in the conversion of pyruvate to acetyl-CoA and in the citric acid cycle

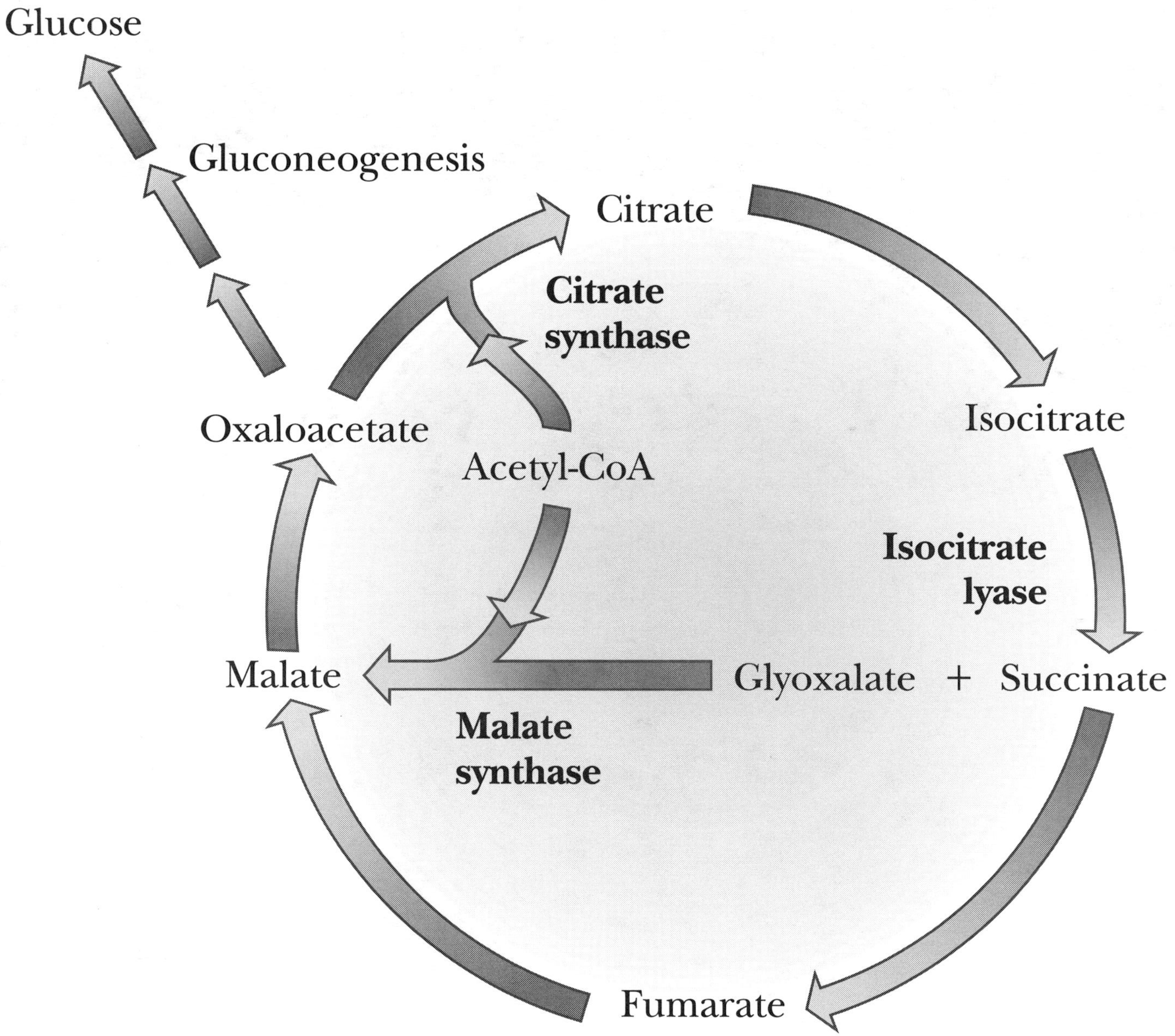

Glucose
Gluconeogenesis
Citrate
Citrate
synthase
Oxaloacetate
Acetyl-CoA
Isocitrate
Isocitrate
lyase
Malate
Glyoxalate + Succinate
Malate
synthase
Fumarate

Figure 16.8 The glyoxylate cycle

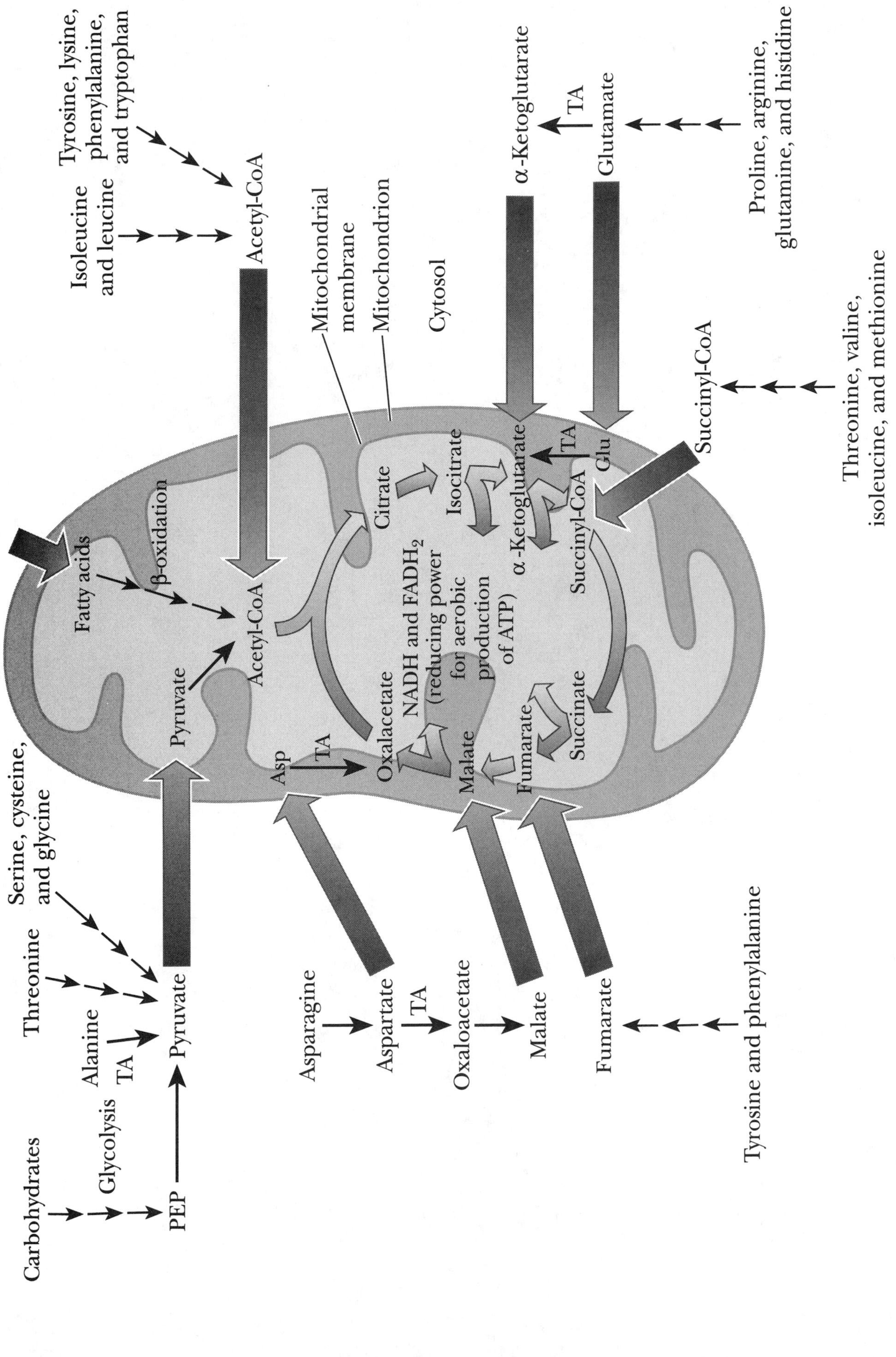

Carbohydrates
Glycolysis
PEP
Threonine
Serine, cysteine, and glycine
Alanine
TA
Pyruvate
Fatty acids
β-oxidation
Pyruvate
Acetyl-CoA
Isoleucine and leucine
Tyrosine, lysine, phenylalanine, and tryptophan
Acetyl-CoA
Asp
TA
Asparagine
Aspartate
TA
Oxaloacetate
Malate
Fumarate
Tyrosine and phenylalanine
Oxalacetate
NADH and FADH$_2$ (reducing power for aerobic production of ATP)
Citrate
Isocitrate
Malate
Fumarate
Succinate
α-Ketoglutarate
Succinyl-CoA
TA
Glu
Mitochondrial membrane
Mitochondrion
Cytosol
α-Ketoglutarate
TA
Glutamate
Proline, arginine, glutamine, and histidine
Succinyl-CoA
Threonine, valine, isoleucine, and methionine

Figure 16.9 A summary of catabolism

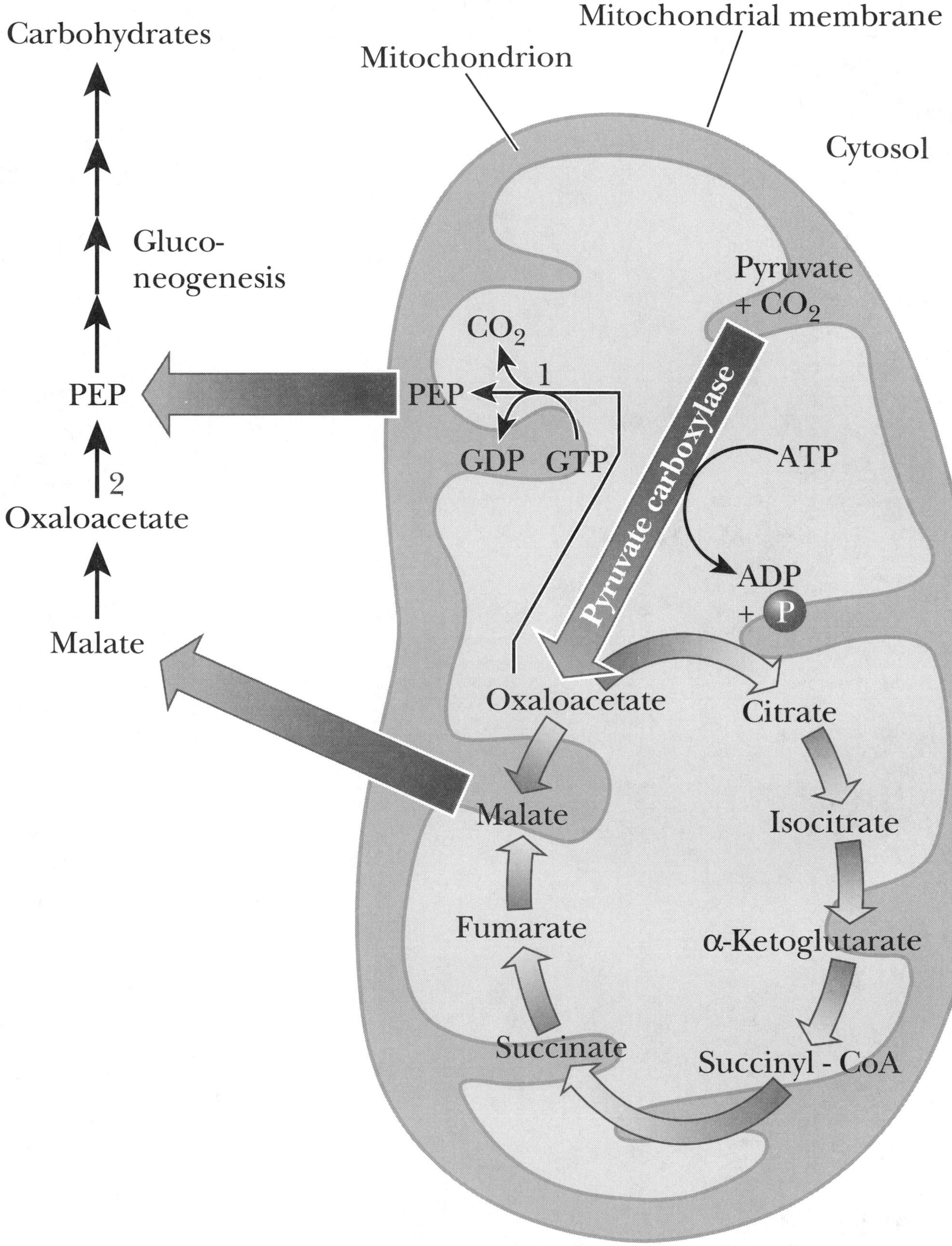

Figure 16.11 Transfer of the starting materials if gluco-neogenesis from the mitochondrion to the cytosol

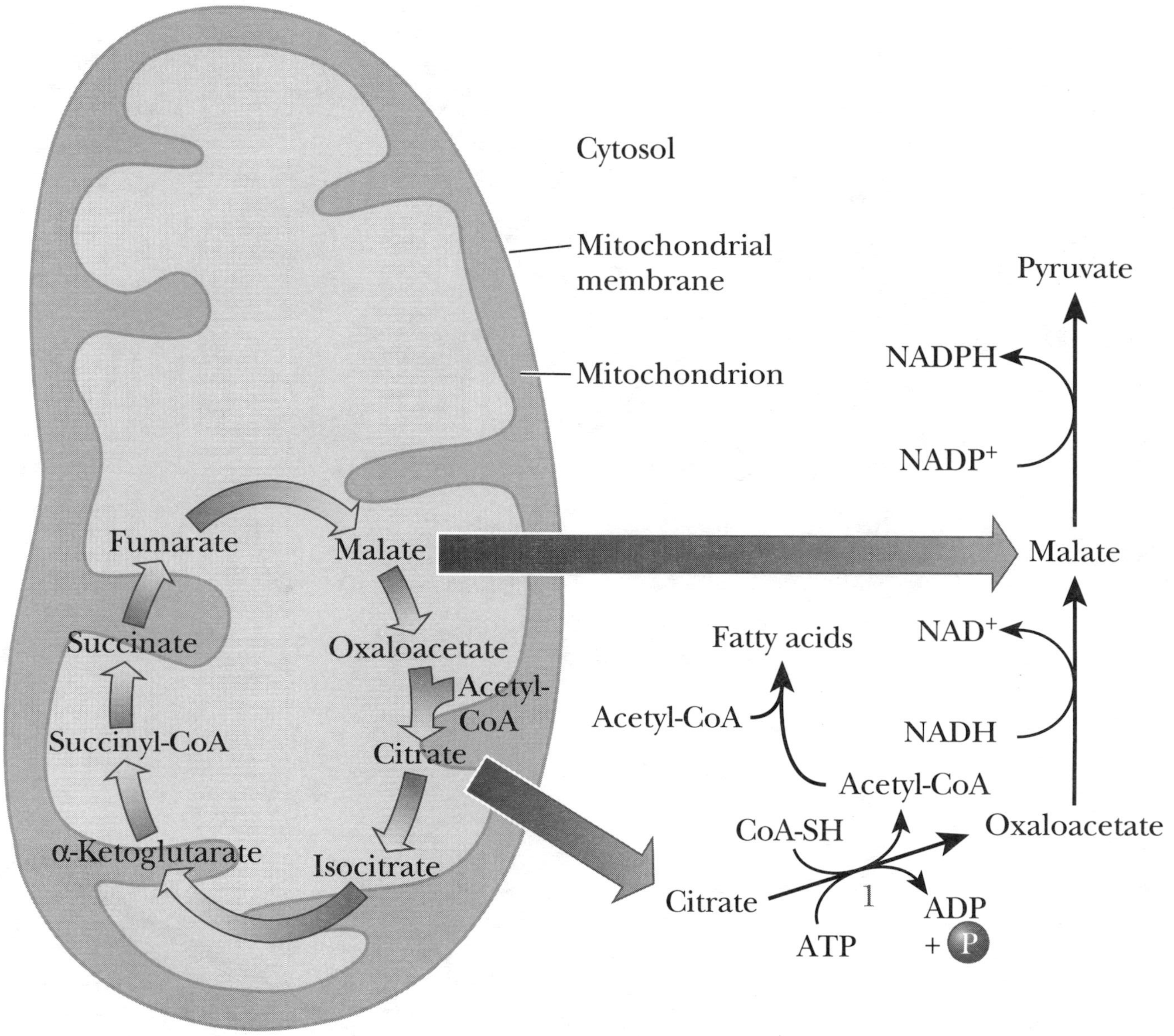

Figure 16.12 Transfer of the starting materials of lipid anabolism from the mitochondrion to the cytosol

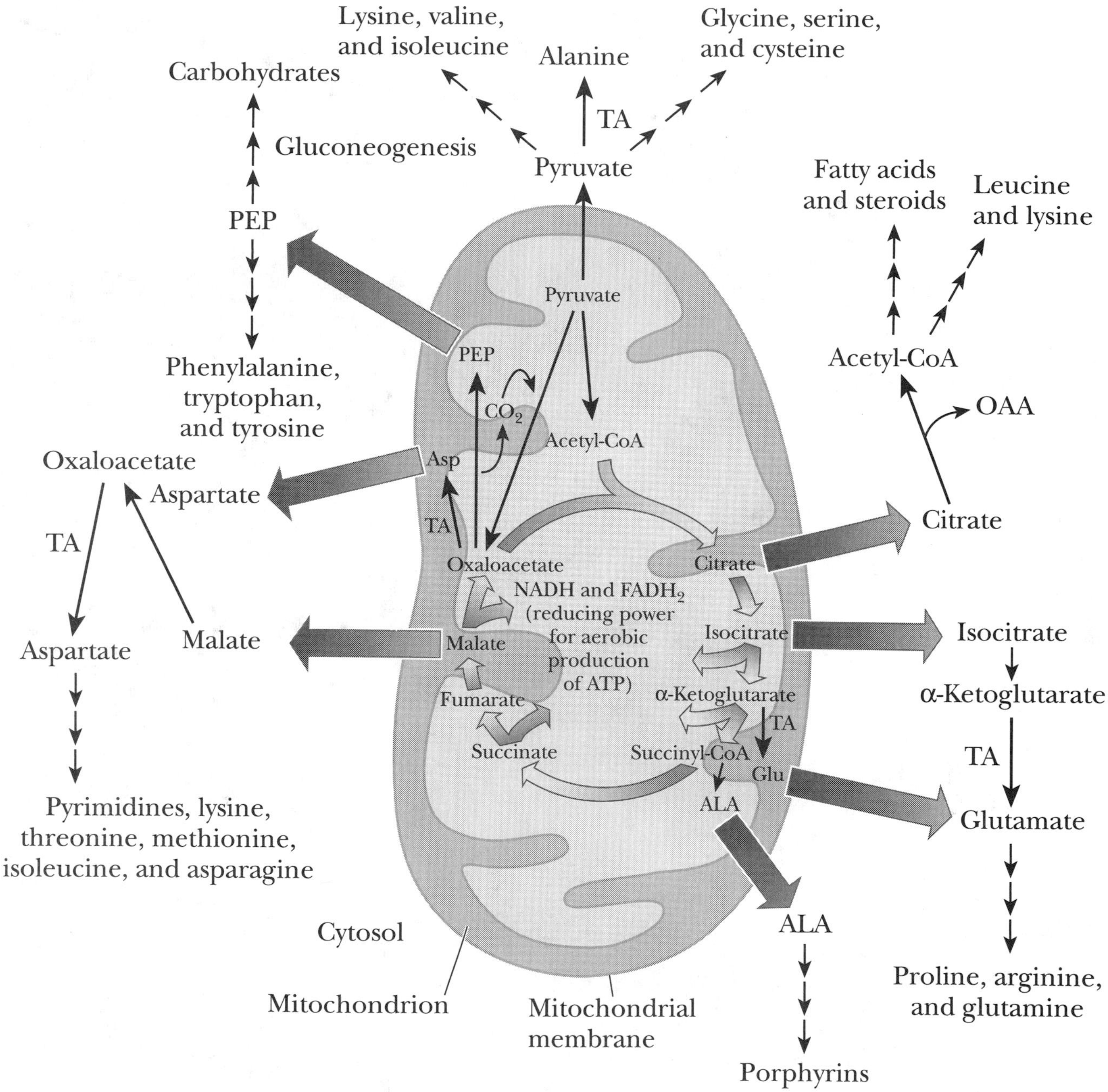

Lysine, valine, and isoleucine
Alanine
Glycine, serine, and cysteine
Carbohydrates
Gluconeogenesis
TA
Pyruvate
Fatty acids and steroids
Leucine and lysine
PEP
Pyruvate
PEP
Acetyl-CoA
Phenylalanine, tryptophan, and tyrosine
CO_2
OAA
Acetyl-CoA
Oxaloacetate
Asp
Aspartate
TA
Citrate
TA
Oxaloacetate
Citrate
NADH and $FADH_2$ (reducing power for aerobic production of ATP)
Aspartate
Malate
Malate
Isocitrate
Isocitrate
α-Ketoglutarate
Fumarate
α-Ketoglutarate
TA
Succinate
Succinyl-CoA
TA
Glu
ALA
Glutamate
Pyrimidines, lysine, threonine, methionine, isoleucine, and asparagine
ALA
Cytosol
Proline, arginine, and glutamine
Mitochondrion
Mitochondrial membrane
Porphyrins

Figure 16.14 A summary of anabolism

TABLE 16.2 The Energetics of Conversion of Pyruvate to Carbon Dioxide

Step	Reaction	$\Delta G°'$ kJ mol^{-1}	$\Delta G°'$ kcal mol^{-1}
	Pyruvate + CoA-SH + NAD^+ → Acetyl-CoA + NADH + CO_2	−33.4	−8.0
1	Acetyl-CoA + Oxaloacetate + H_2O → Citrate + CoA-SH + H^+	−32.2	−7.7
2	Citrate → Isocitrate	+6.3	+1.5
3	Isocitrate + NAD^+ → α-Ketoglutarate + NADH + CO_2 + H^+	−7.1	−1.7
4	α-Ketoglutarate + $NADH^+$ + CoA-SH → Succinyl-CoA + NADH + CO_2 + H^+	−33.4	−8.0
5	Succinyl-CoA + GDP + P_i → Succinate + GTP + CoA-SH	−3.3	−0.8
6	Succinate + FAD → Fumarate + $FADH_2$	≈0	≈0
7	Fumarate + H_2O → L-Malate	−3.8	−0.9
8	L-Malate + NAD^+ → Oxaloacetate + NADH + H^+	+29.2	+7.0
Overall:	Pyruvate + 4 $NADH^+$ + FAD + GDP + P_i + 2 H_2O → 3 CO_2 + 4 NADH + $FADH_2$ + GTP + 4 H^+	−77.7	−18.6

Table 16.2 The energetics of conversion of pyruvate to carbon dioxide

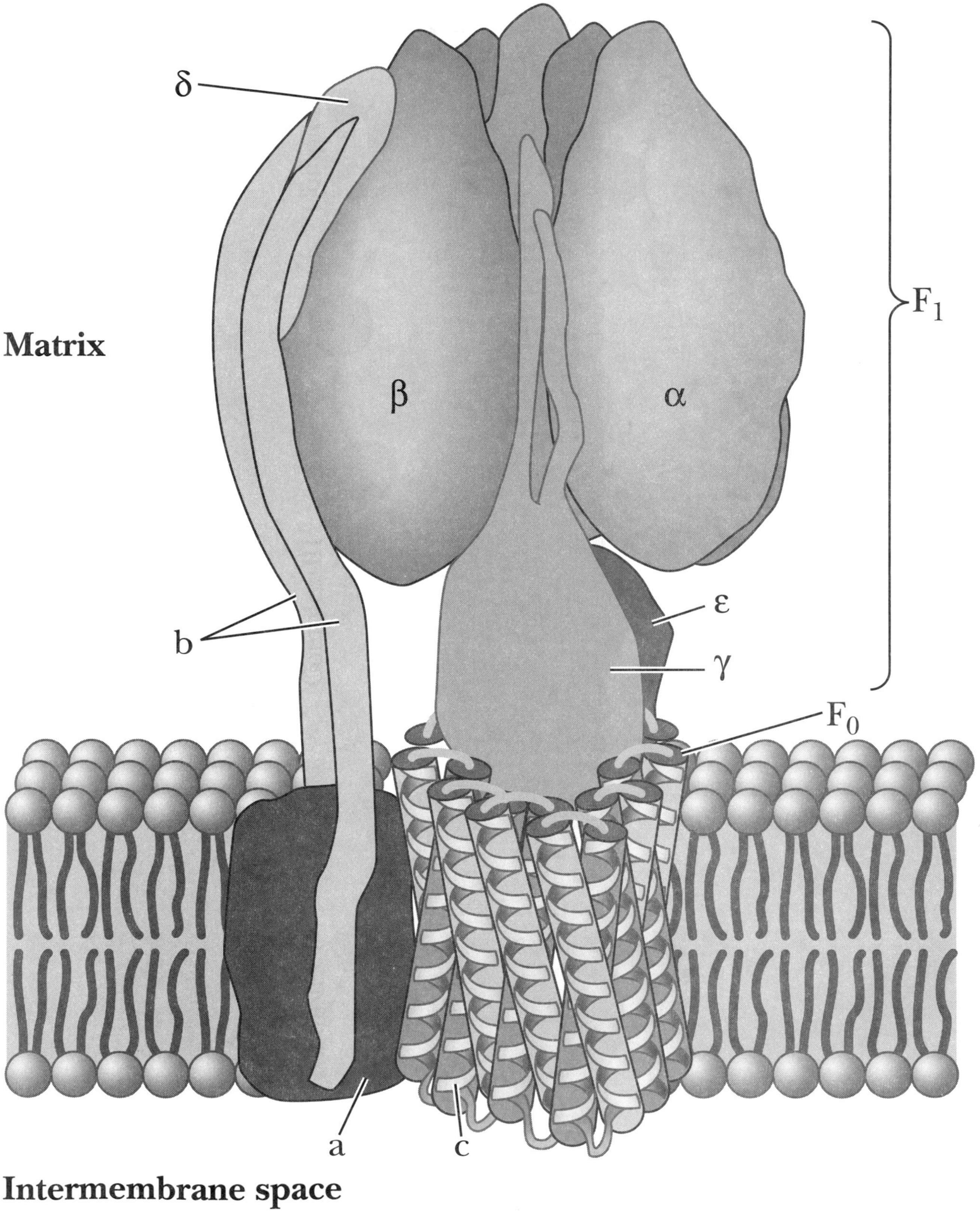

Figure 17.11 A model for the components of ATP synthase

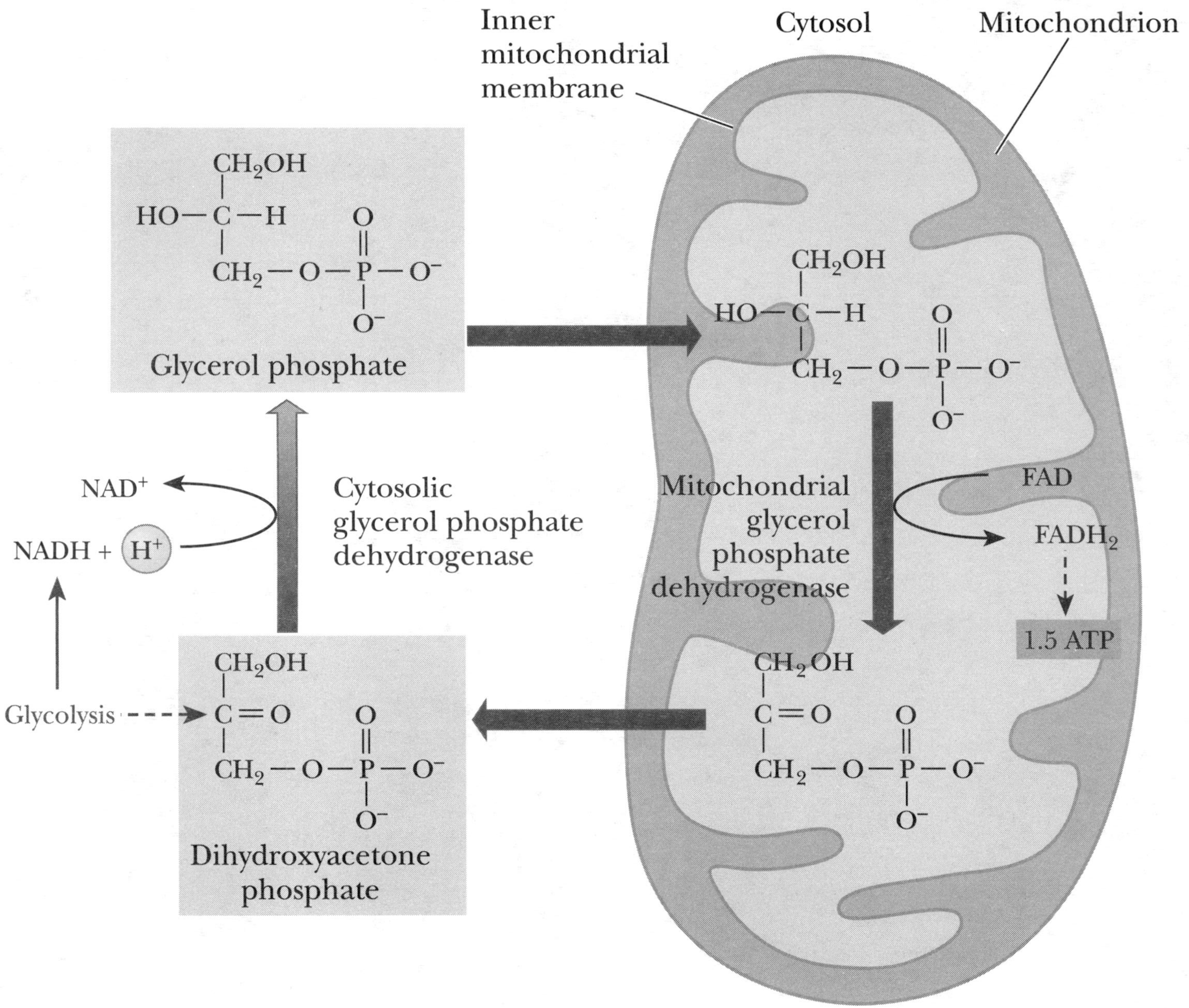

Inner mitochondrial membrane
Cytosol
Mitochondrion
Glycerol phosphate
Cytosolic glycerol phosphate dehydrogenase
NAD+
NADH + H+
Glycolysis
Dihydroxyacetone phosphate
Mitochondrial glycerol phosphate dehydrogenase
FAD
FADH2
1.5 ATP

Figure 17.21 The glycerol phosphate shuttle

Inner mitochondrial membrane
Cytosol
Mitochondrion
α-Ketoglutarate
Aspartate
L-Aspartate
L-Aspartate
α-Ketoglutarate
α-Ketoglutarate
Cytosolic aspartate aminotransferase
Mitochondrial aspartate aminotransferase
L-Glutamate
L-Glutamate
Glycolysis
Oxaloacetate
Oxaloacetate
2.5 ATP
Cytosolic malate dehydrogenase
Mitochondrial malate dehydrogenase (citric acid cycle)
NADH + H+
NAD+
NADH
+ H+
NAD+
L-Malate
L-Malate
Oxaloacetate
Glutamate
Malate

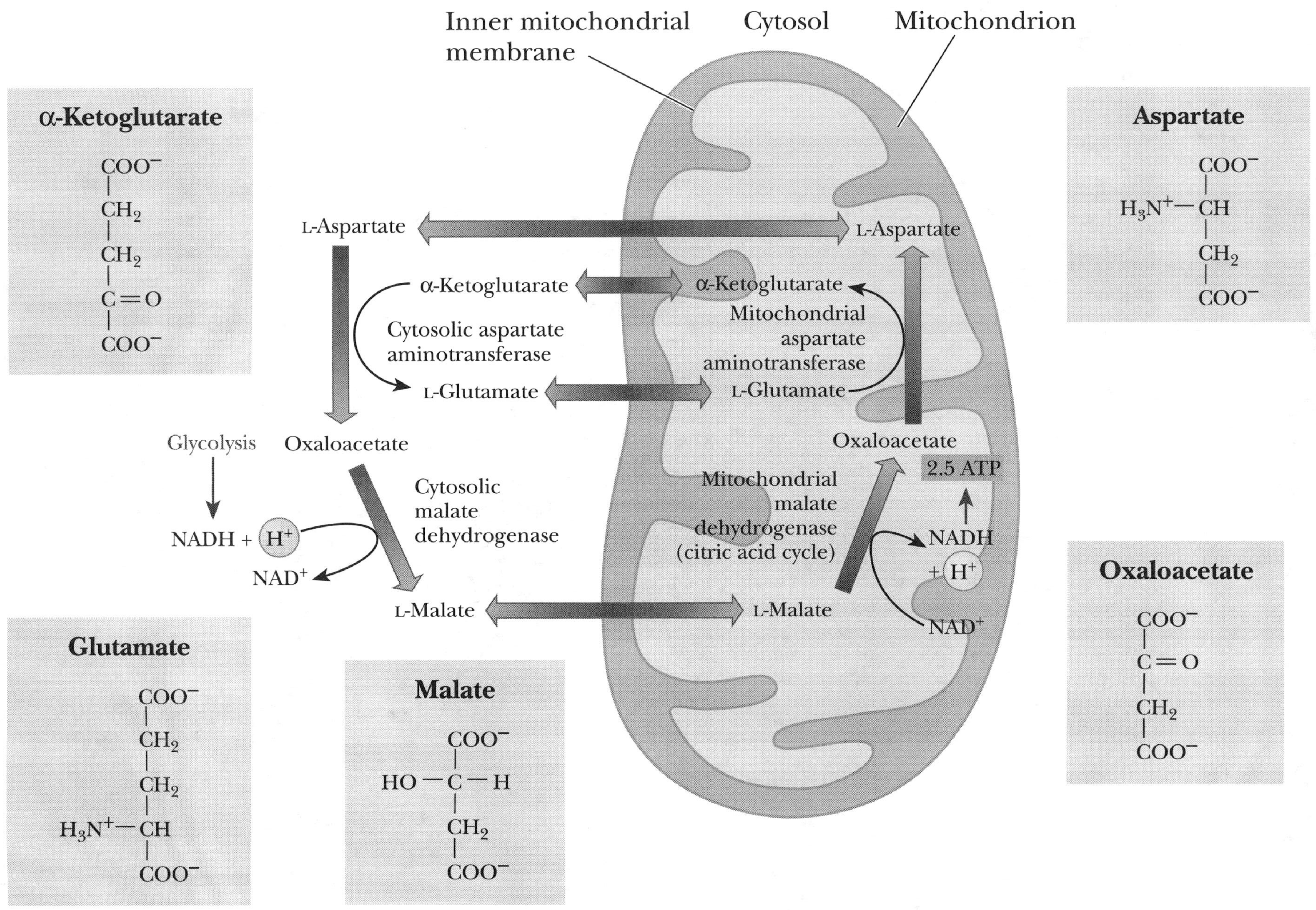

Figure 17.22 The malate-aspartate shuttle

TABLE 17.1 Standard Reduction Potentials for Several Biological Reduction Half-Reactions

Reduction Half-Reaction	$E°'(V)$
$\frac{1}{2} O_2 + 2\ H^+ + 2e^- \rightarrow H_2O$	0.816
$Fe^{3+} + e^- \rightarrow Fe^{2+}$	0.771
Cyt $a_3(Fe^{3+}) + e^- \rightarrow$ Cyt $a_3(Fe^{2+})$	0.350
Cyt $a(Fe^{3+}) + e^- \rightarrow$ Cyt $a(Fe^{2+})$	0.290
Cyt $c(Fe^{3+}) + e^- \rightarrow$ Cyt $c(Fe^{2+})$	0.254
Cyt $c_1(Fe^{3+}) + e^- \rightarrow$ Cyt $c_1(Fe^{2+})$	0.220
CoQH $+ H^+ + e^- \rightarrow CoQH_2$	0.190
CoQ $+ 2\ H^+ + 2\ e^- \rightarrow CoQH_2$	0.060
Cyt $b_H(Fe^{3+}) + e^- \rightarrow$ Cyt $b_H(Fe^{2+})$	0.050
Fumarate $+ 2\ H^+ + 2\ e^- \rightarrow$ Succinate	0.031
CoQ $+ H^+ + e^- \rightarrow$ CoQH	0.030
[FAD] $+ 2\ H^+ + 2\ e^- \rightarrow [FADH_2]$	0.003–0.091*
Cyt $b_L(Fe^{3+}) + e^- \rightarrow$ Cyt $b_L(Fe^{2+})$	−0.100
Oxaloacetate $+ 2\ H^+ + 2\ e^- \rightarrow$ Malate	−0.166
Pyruvate $+ 2\ H^+ + 2\ e^- \rightarrow$ Lactate	−0.185
Acetaldehyde $+ 2\ H^+ + 2\ e^- \rightarrow$ Ethanol	−0.197
FMN $+ 2\ H^+ + 2\ e^- \rightarrow FMNH_2$	−0.219
FAD $+ 2\ H^+ + 2\ e^- \rightarrow FADH_2$	−0.219
1,3-*Bis*phosphoglycerate $+ 2\ H^+ + 2\ e^- \rightarrow$ Glyceraldehyde-3-	−0.290
phosphate $+ P_i$	−0.320
$NAD^+ + 2\ H^+ + 2\ e^- \rightarrow NADH + H^+$	−0.320
NADP+ $+ 2\ H^+ + 2\ e^- \rightarrow NADPH + H^+$	−0.380
α-Ketoglutarate $+ CO_2 + 2\ H^+ + 2\ e^- \rightarrow$ Isocitrate	−0.670
Succinate $+ CO_2 + 2\ H^+ + 2\ e^- \rightarrow$ α-Ketoglutarate $+ H_2O$	

*Typical values for reduction of bound FAD in flavoproteins such as succinate dehydrogenase.

Note that we have shown a number of components of the electron transport chain individually. We will see them again as part of complexes. We have also included values for a number of reactions we saw in earlier chapters.

Table 17.1 Standard reduction potentials for several biological reduction half-reactions

TABLE 17.2 The Energetics of Electron Transport Reactions

Reaction	$\Delta G^{\circ\prime}$ kJ (mol NADH)$^{-1}$	$\Delta G^{\circ\prime}$ kcal (mol/NADH)$^{-1}$
$NADH + H^+ + E—FMN \rightarrow NAD^+ + E—FMNH_2$	−38.6	−9.2
$E—FMNH_2 + CoQ \rightarrow E—FMN + CoQH_2$	−42.5	−10.2
$CoQH_2 + 2\ Cyt\ b[Fe(III)] \rightarrow CoQ + 2H^+ + 2\ Cyt\ b[Fe(II)]$	+11.6	+2.8
$2\ Cyt\ b[Fe(II)] + 2\ Cyt\ c_1[Fe(III)] \rightarrow 2\ Cyt\ c_1[Fe(II)] + 2\ Cyt\ b[Fe(III)]$	−34.7	−8.3
$2\ Cyt\ c_1[Fe(II)] + 2\ Cyt\ c[Fe(III)] \rightarrow 2\ Cyt\ c[Fe(II)] + 2\ Cyt\ c_1[Fe(III)]$	−5.8	−1.4
$2\ Cyt\ c[Fe(II)] + 2\ Cyt\ (aa_3)\ [Fe(III)] \rightarrow 2\ Cyt\ (aa_3)\ [Fe(II)] + 2\ Cyt\ c[Fe(III)]$	−7.7	−1.8
$2\ Cyt\ (aa_3)[Fe(II)] + \frac{1}{2} O_2 + 2\ H^+ \rightarrow 2\ Cyt\ (aa_3)[Fe(III)] + H_2O$	−102.3	−24.5
Overall reaction: $NADH + H^+ + \frac{1}{2} O_2 \rightarrow NAD^+ + H_2O$	−220	−52.6

Table 17.2 Energetics of electron transport reactions

TABLE 17.3 Yield of ATP from Glucose Oxidation

Pathway	ATP Yield per Glucose: *Glycerol–Phosphate Shuttle*	ATP Yield per Glucose: *Malate–Aspartate Shuttle*	NADH	$FADH_2$
Glycolysis: glucose to pyruvate (cytosol)				
Phosphorylation of glucose	−1	−1		
Phosphorylation of fructose-6-phosphate	−1	−1		
Dephosphorylation of 2 molecules of 1,3-BPG	+2	+2		
Dephosphorylation of 2 molecules of PEP	+2	+2		
Oxidation of 2 molecules of glyceraldehyde-3-phosphate yields 2 NADH			+2	
Pyruvate conversion to acetyl-CoA (mitochondria)				
2 NADH produced			+2	
Citric acid cycle (mitochondria)				
2 molecules of GTP from 2 molecules of succinyl-CoA	+2	2		
Oxidation of 2 molecules each of isocitrate, α-ketoglutarate, and malate yields 6 NADH			+6	
Oxidation of 2 molecules of succinate yields 2 $FADH_2$				+2
Oxidative phosphorylation (mitochondria)				
2 NADH from glycolysis yield 1.5 ATP each if NADH is oxidized by glycerol–phosphate shuttle; 2.5 ATP by malate–aspartate shuttle	+3	+5	−2	
Oxidative decarboxylation of 2 pyruvate to 2 acetyl-CoA: 2 NADH produce 2.5 ATP each	+5	+5	−2	
2 $FADH_2$ from each citric acid cycle produce 1.5 ATP each	+3	+3		−2
6 NADH from citric acid cycle produce 2.5 ATP each	+15	+15	−6	
Net Yield	+30	+32	0	0

(*Note:* These P/O ratios of 2.5 and 1.5 for mitochondrial oxidation of NADH and $FADH_2$ are "consensus values." Since they may not reflect actual values and since these ratios may change depending on metabolic conditions, these estimates of ATP yield from glucose oxidation are approximate.)

Table 17.3 Yield of ATP from glucose oxidation

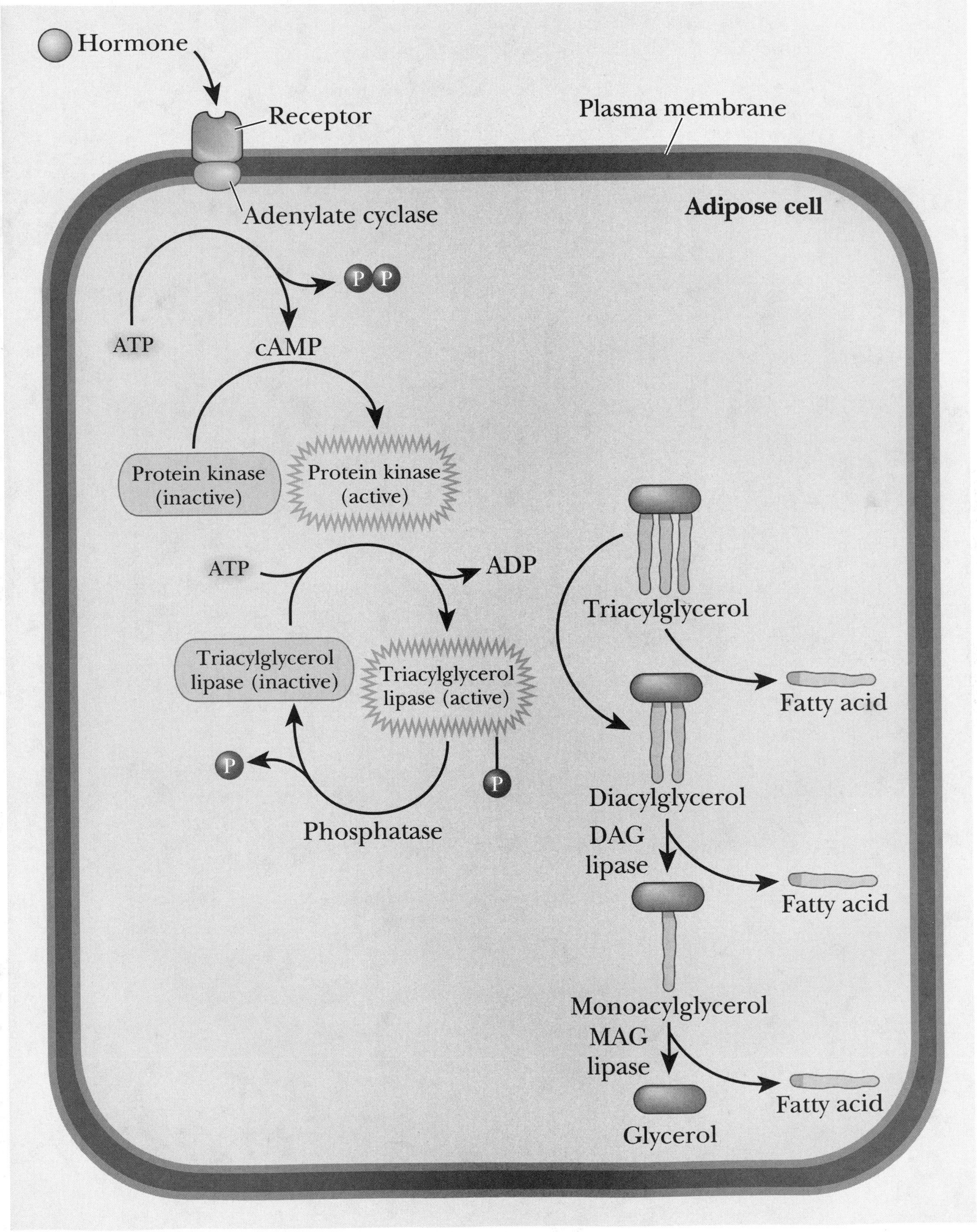

Figure 18.3 Liberation of fatty acids from triacylglycerols

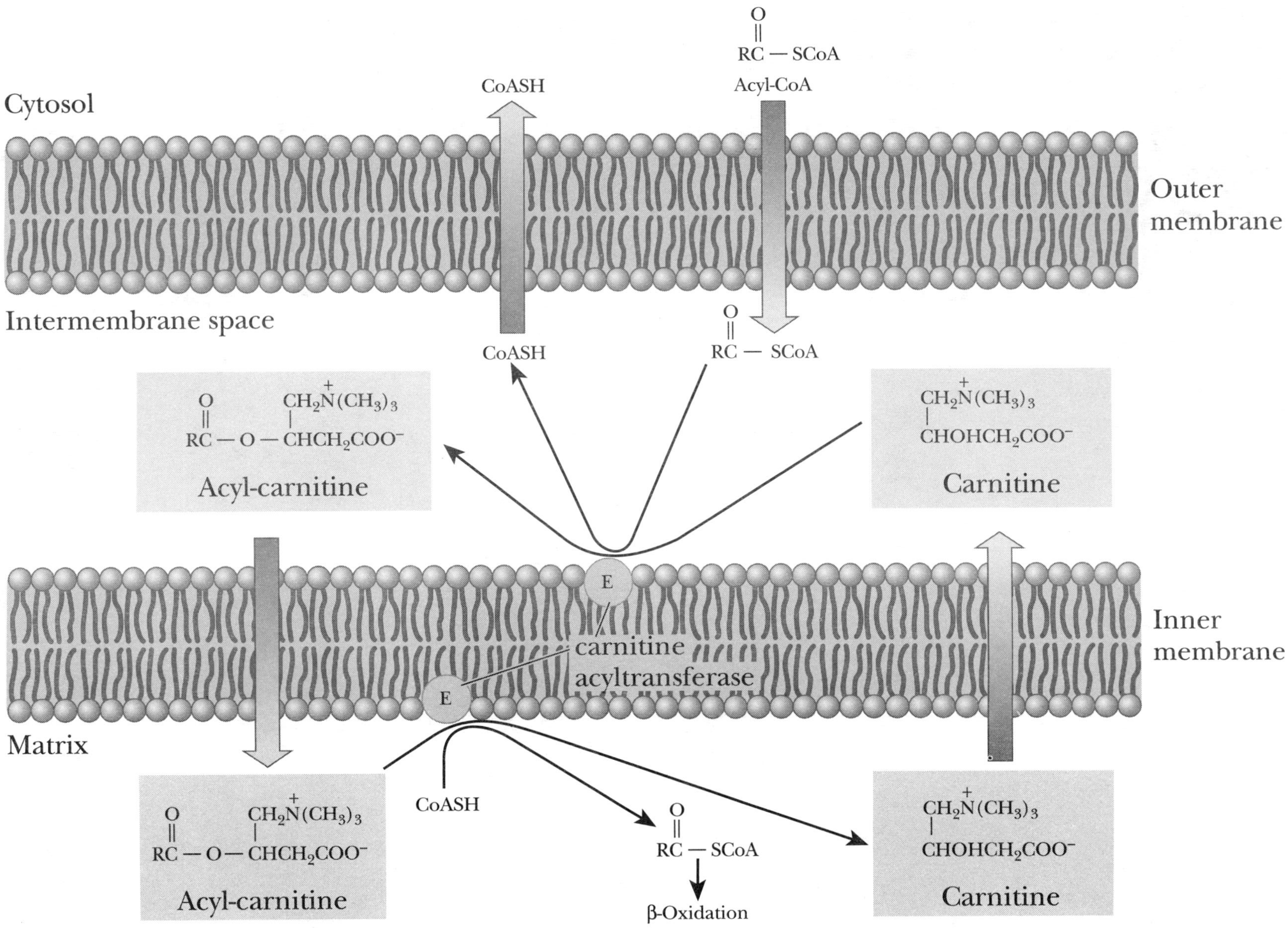

Figure 18.5 The role of carnitine in the transfer of acyl groups to the mitochondrial matrix

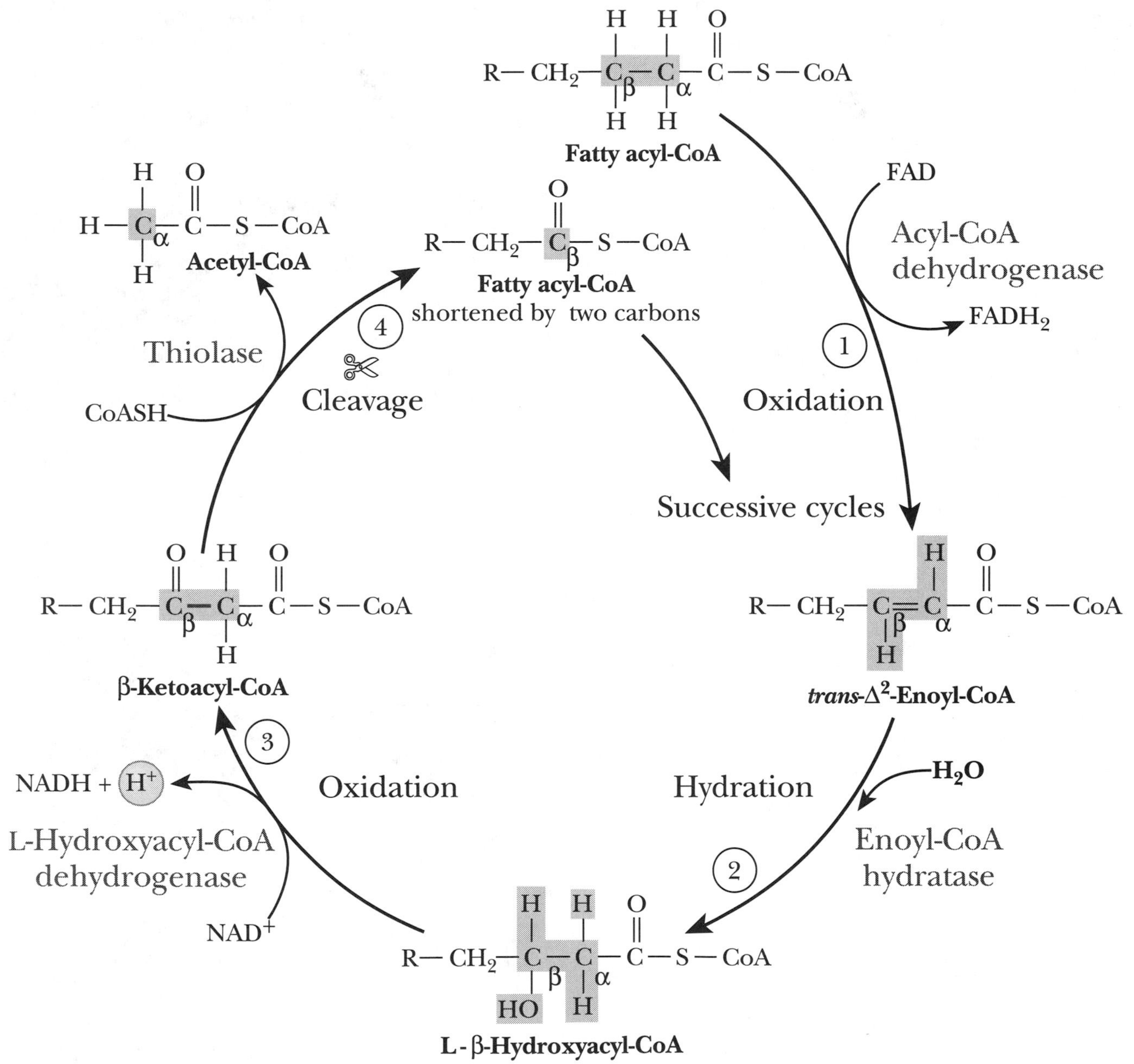

Figure 18.6 The β-oxidation of saturated fatty acids

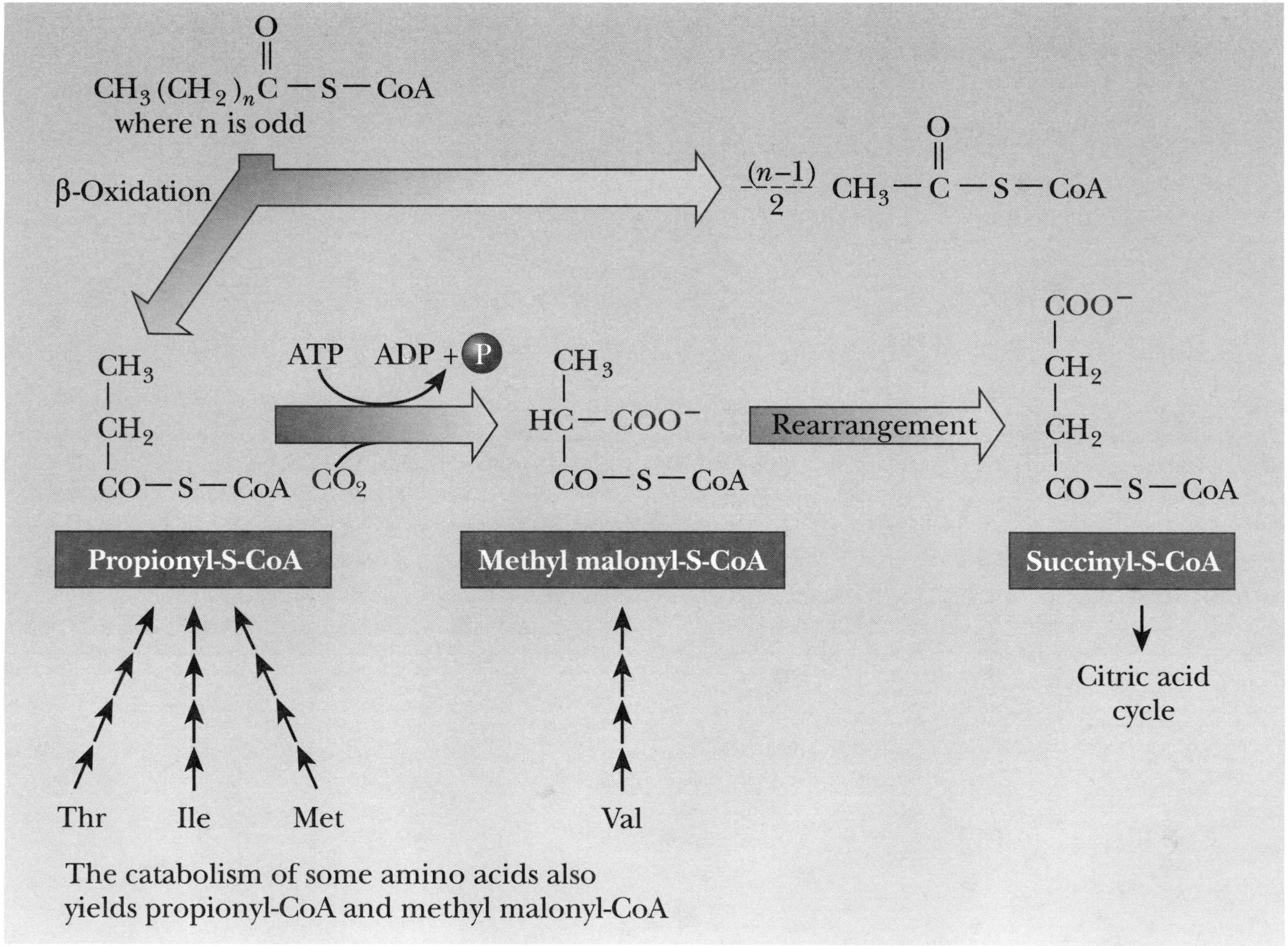

Figure 18.8 The oxidation of fatty acid containing an odd number of carbon atoms

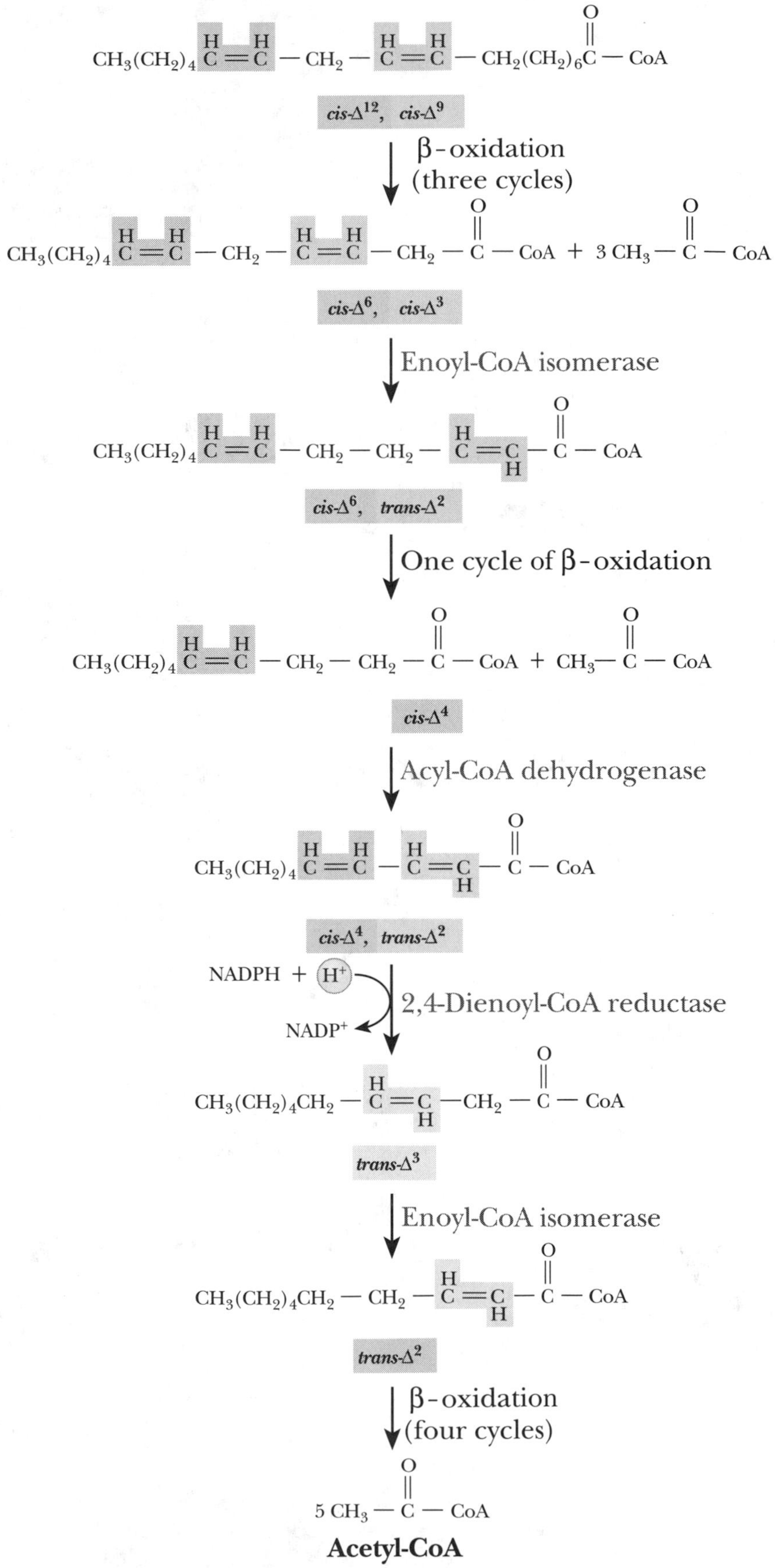

Figure 18.10 The oxidation pathway for polyunsaturated fatty acids

Step 1. Priming of the system by acetyl-CoA

a. ACP-Acyltransferase reaction

$$H_3C-\overset{\overset{\displaystyle O}{\|}}{C}-S-CoA + ACP-SH \rightleftharpoons H_3C-\overset{\overset{\displaystyle O}{\|}}{C}-S-ACP + CoA-SH$$

Acetyl-CoA   **Acetyl-ACP**

b. Transfer to β-ketoacyl-ACP synthase

$$H_3C-\overset{\overset{\displaystyle O}{\|}}{C}-S-ACP + Synthase-SH \rightleftharpoons H_3C-\overset{\overset{\displaystyle O}{\|}}{C}-S-Synthase + ACP-SH$$

Acetyl-ACP   **Acetyl-synthase**

Step 2. ACP-malonyltransferase reaction (malonyl transfer to system)

$$^-OOC-CH_2-\overset{\overset{\displaystyle O}{\|}}{C}-S-CoA + ACP-SH \rightleftharpoons {}^-OOC-CH_2-\overset{\overset{\displaystyle O}{\|}}{C}-S-ACP + CoA-SH$$

Malonyl-CoA   **Malonyl-ACP**

Step 3. β-Ketoacyl-ACP synthase reaction (condensation)

$$H_3C-\overset{\overset{\displaystyle O}{\|}}{C}-S-Synthase + {}^-OOC-CH_2-\overset{\overset{\displaystyle O}{\|}}{C}-S-ACP \rightleftharpoons H_3C-\overset{\overset{\displaystyle O}{\|}}{C}-CH_2-\overset{\overset{\displaystyle O}{\|}}{C}-S-ACP + CO_2 + Synthase-SH$$

Acetyl-synthase   **Malonyl-ACP**   **Acetoacetyl-ACP**

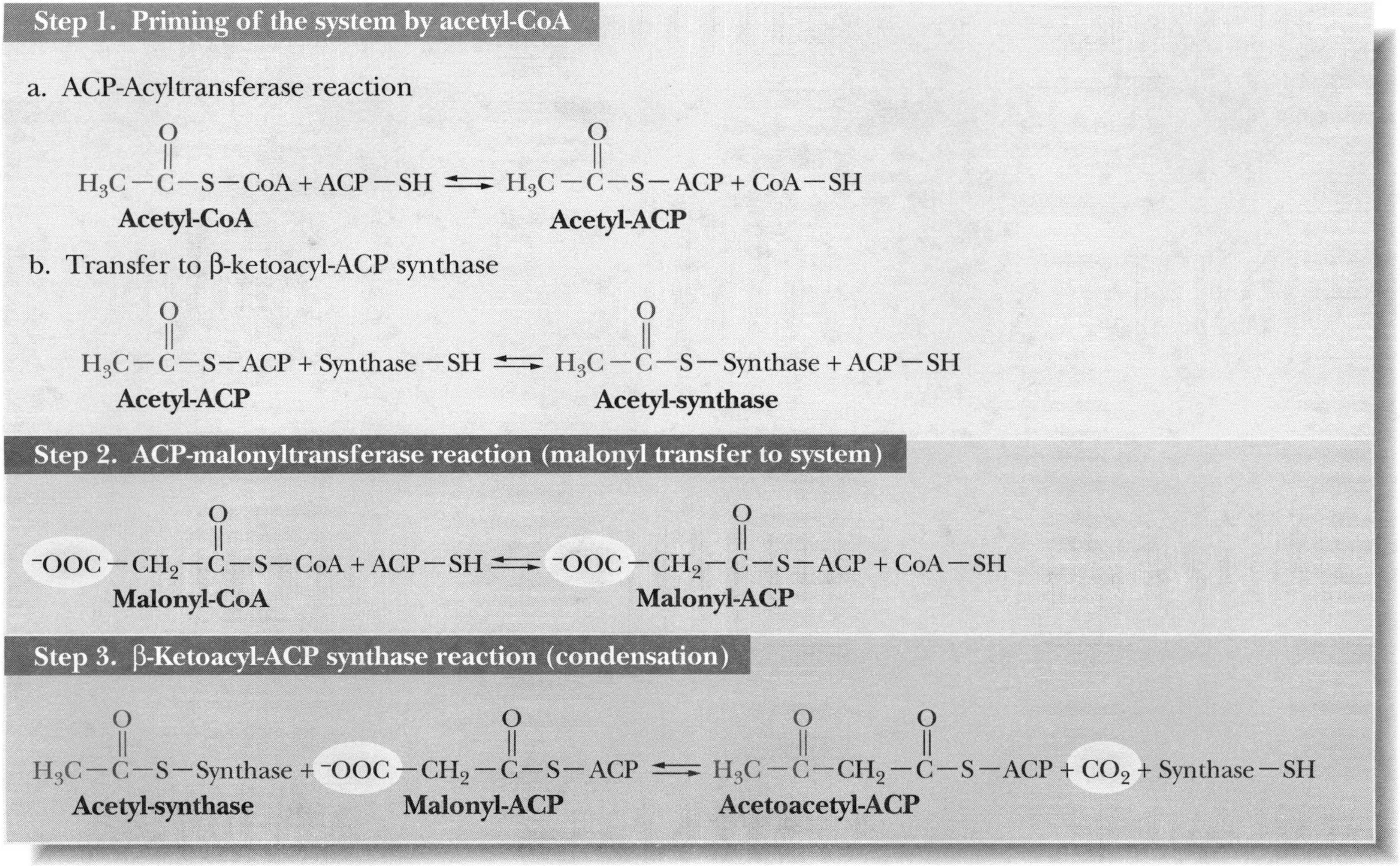

Figure 18.15 The first cycle of palmitate synthesis

Step 4. β-Ketoacyl-ACP reductase reaction (first reduction)

$$H_3C-C-CH_2-C-S-ACP + NADPH + H^+ \rightleftharpoons H_3C-CH-CH_2-C-S-ACP + NADP^+$$

Acetoacetyl-ACP   D-β-Hydroxybutyryl-ACP

Step 5. β-Hydroxyacyl-ACP dehydratase (dehydration)

$$H_3C-CH-CH_2-C-S-ACP \rightleftharpoons H_3C-CH=CH-C-S-ACP + H_2O$$

D-β-Hydroxybutyryl-ACP   Crotonyl-ACP

Step 6. β-Enoyl-ACP reductase (second reduction)

$$H_3C-CH=CH-C-S-ACP + NADPH + H^+ \rightleftharpoons H_3C-CH_2-CH_2-C-S-ACP + NADP^+$$

Crotonyl-ACP   Butyryl-ACP

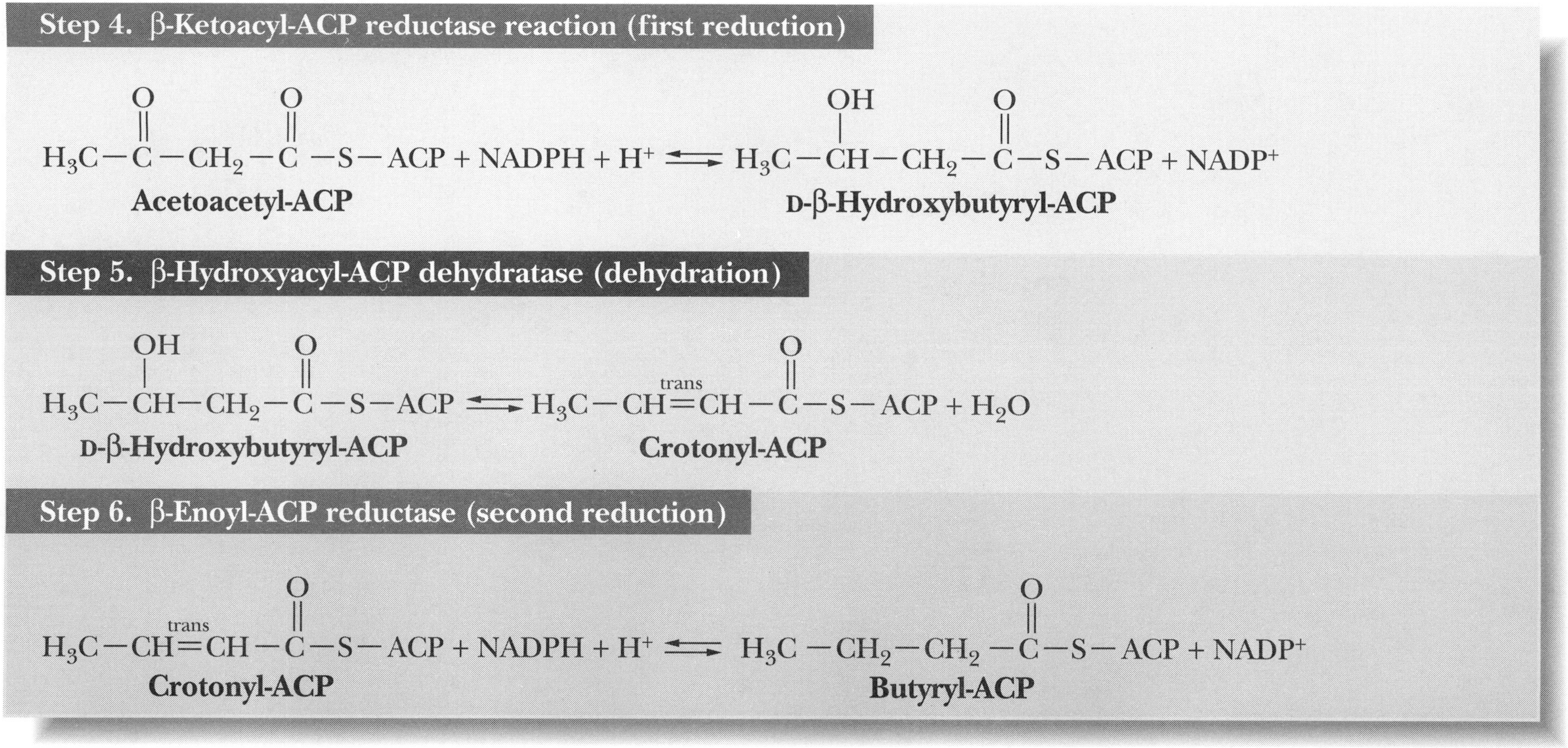

Figure 18.15 (continued) The first cycle of palmitate synthesis

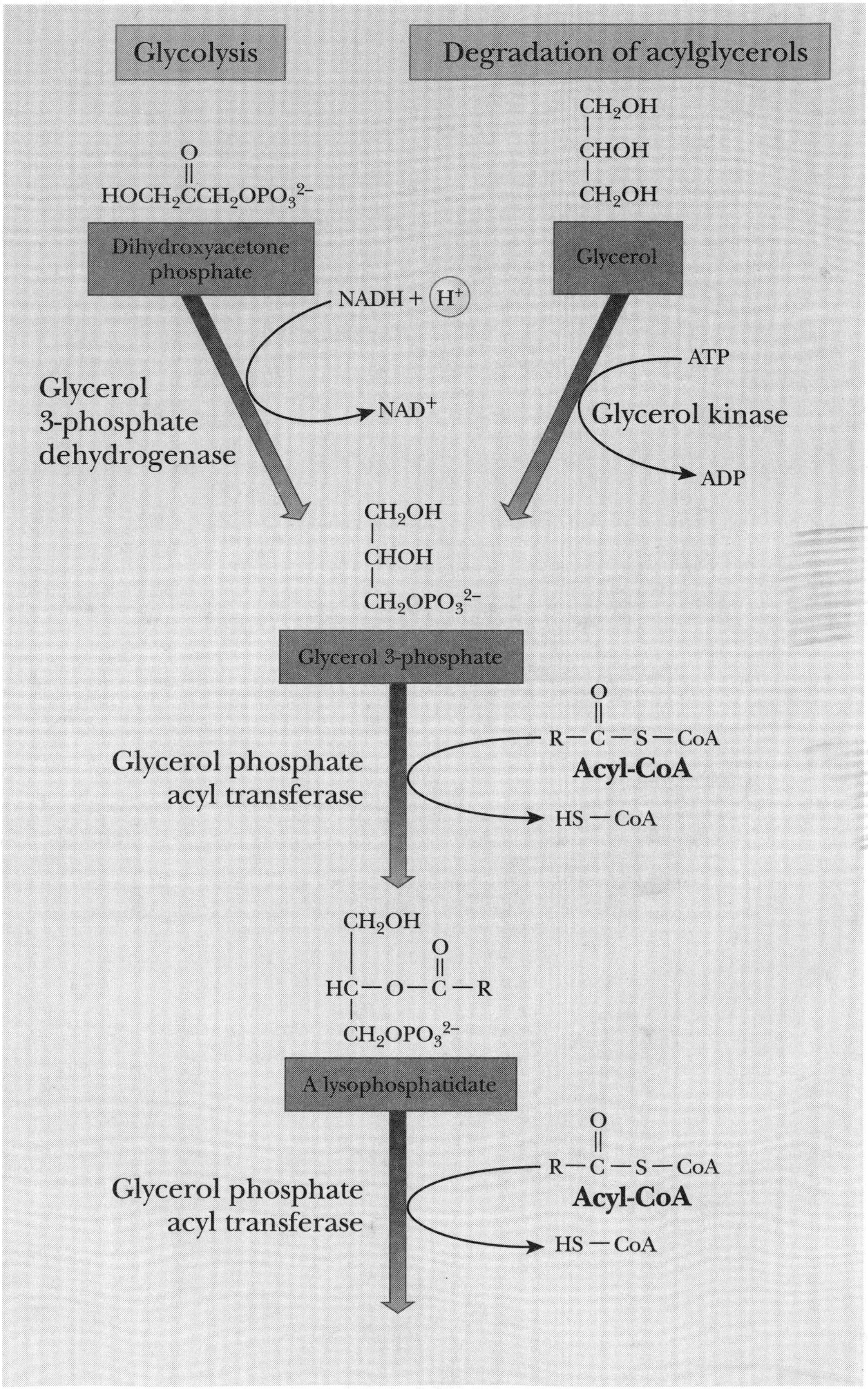

Figure 18.19 (left) Pathways for the biosynthesis of triacylglycerols

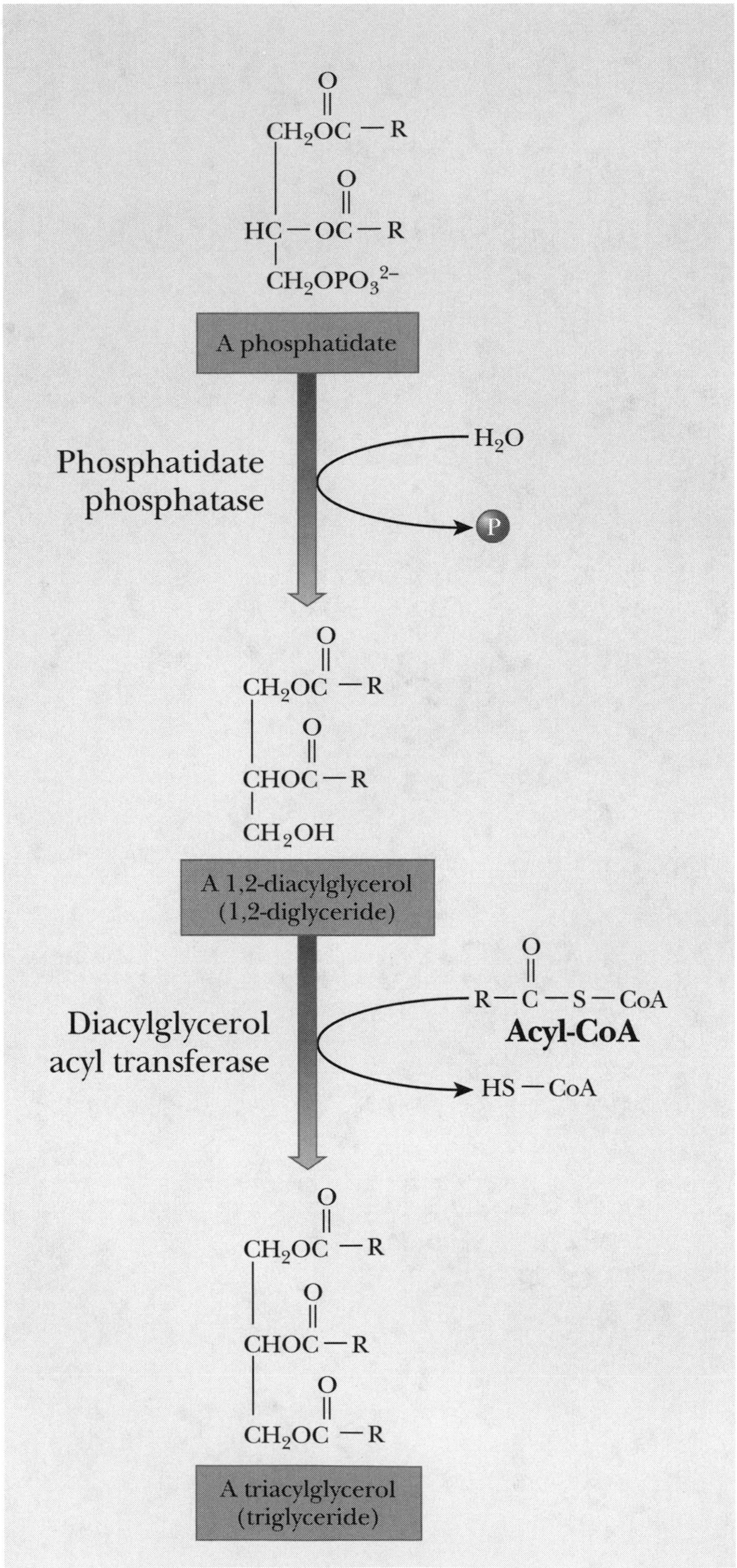

Figure 18.19 (right) Pathways for the biosynthesis of triacylglycerols

TABLE 18.1 The Balance Sheet for Oxidation of One Molecule of Stearic Acid

Reaction	NADH Molecules	$FADH_2$ Molecules	ATP Molecules
1. Stearic acid → Stearyl-CoA (activation step)			−2
2. Stearyl-CoA → 9 acetyl-CoA (8 cycles of β-oxidation)	+8	+8	
3. 9 Acetyl-CoA → 18 CO_2 (citric acid cycle); GDP → GTP (9 molecules)	+27	+9	+9
4. Reoxidation of NADH from β-oxidation cycle	−8		+20
5. Reoxidation of NADH from citric acid cycle	−27		+67.5
6. Reoxidation of $FADH_2$ from β-oxidation cycle		−8	+12
7. Reoxidation of $FADH_2$ from citric acid cycle		−9	+13.5
	0	0	+120

Note that there is no net change in the number of molecules of NADH or $FADH_2$.

Table 18.1 The balance sheet for oxidation of one molecule of stearic acid

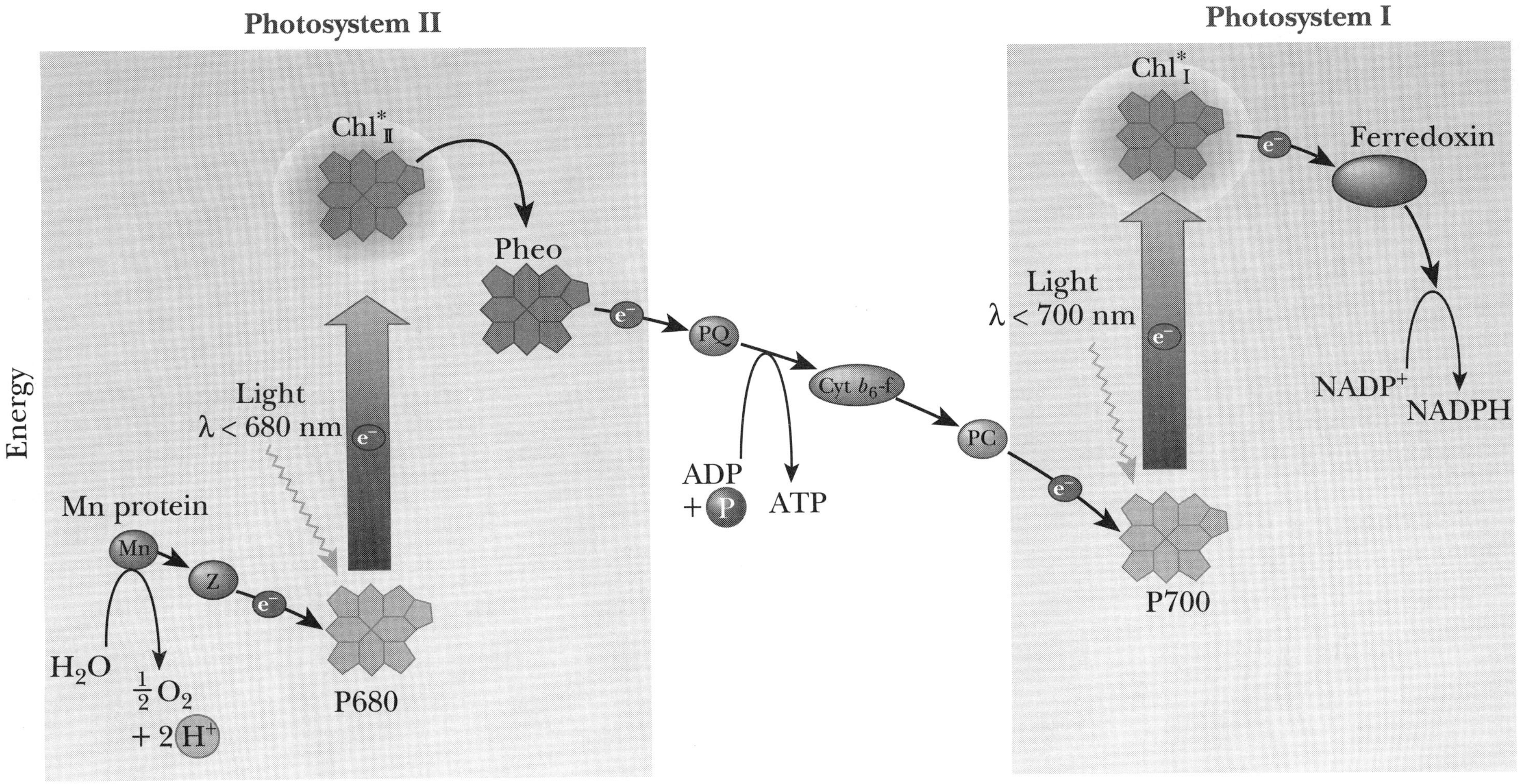

Figure 19.5 Electron flow in Photosystems I and II

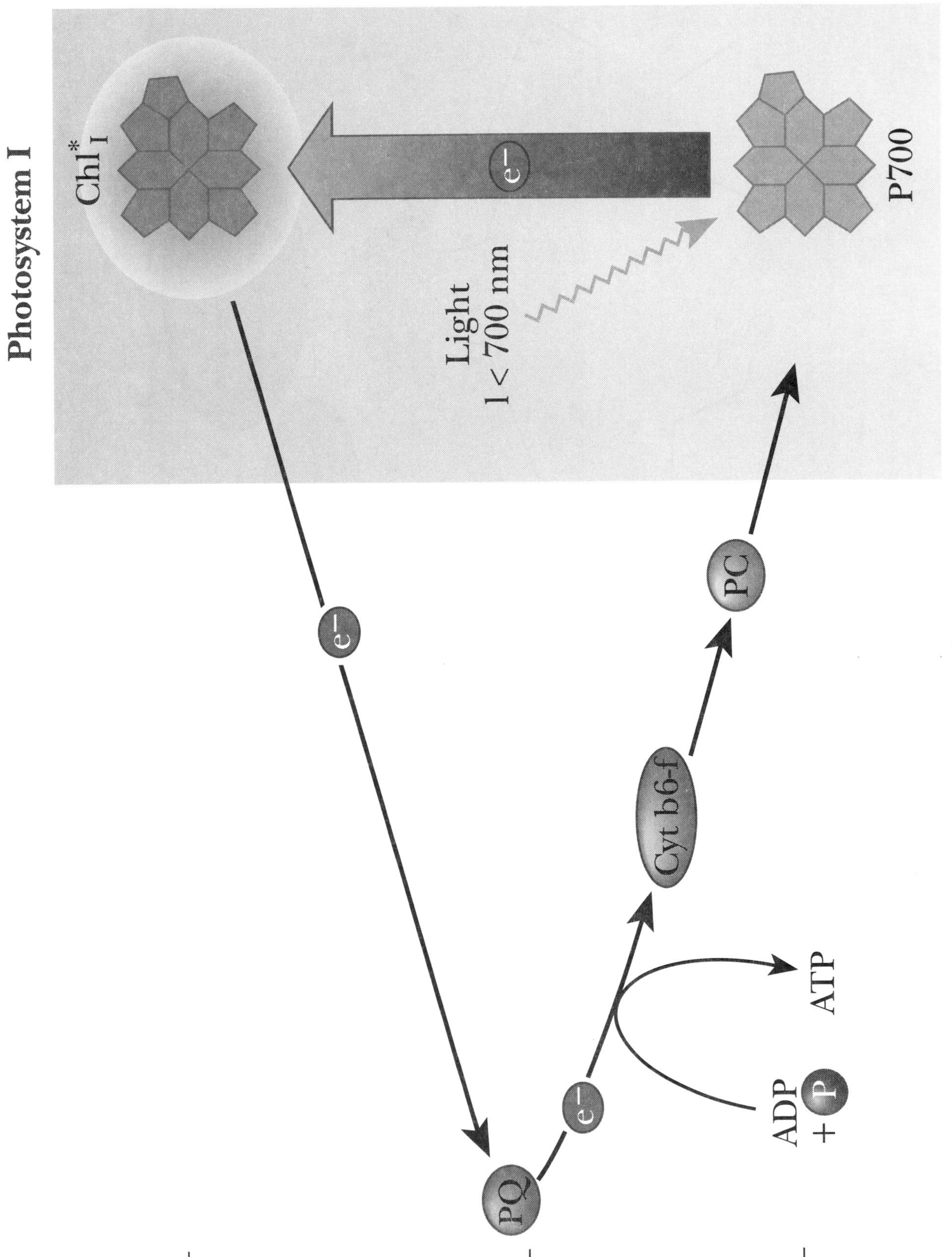

Figure 19.8 Cyclic electron flow coupled to photophosphorylation in Photosystem I

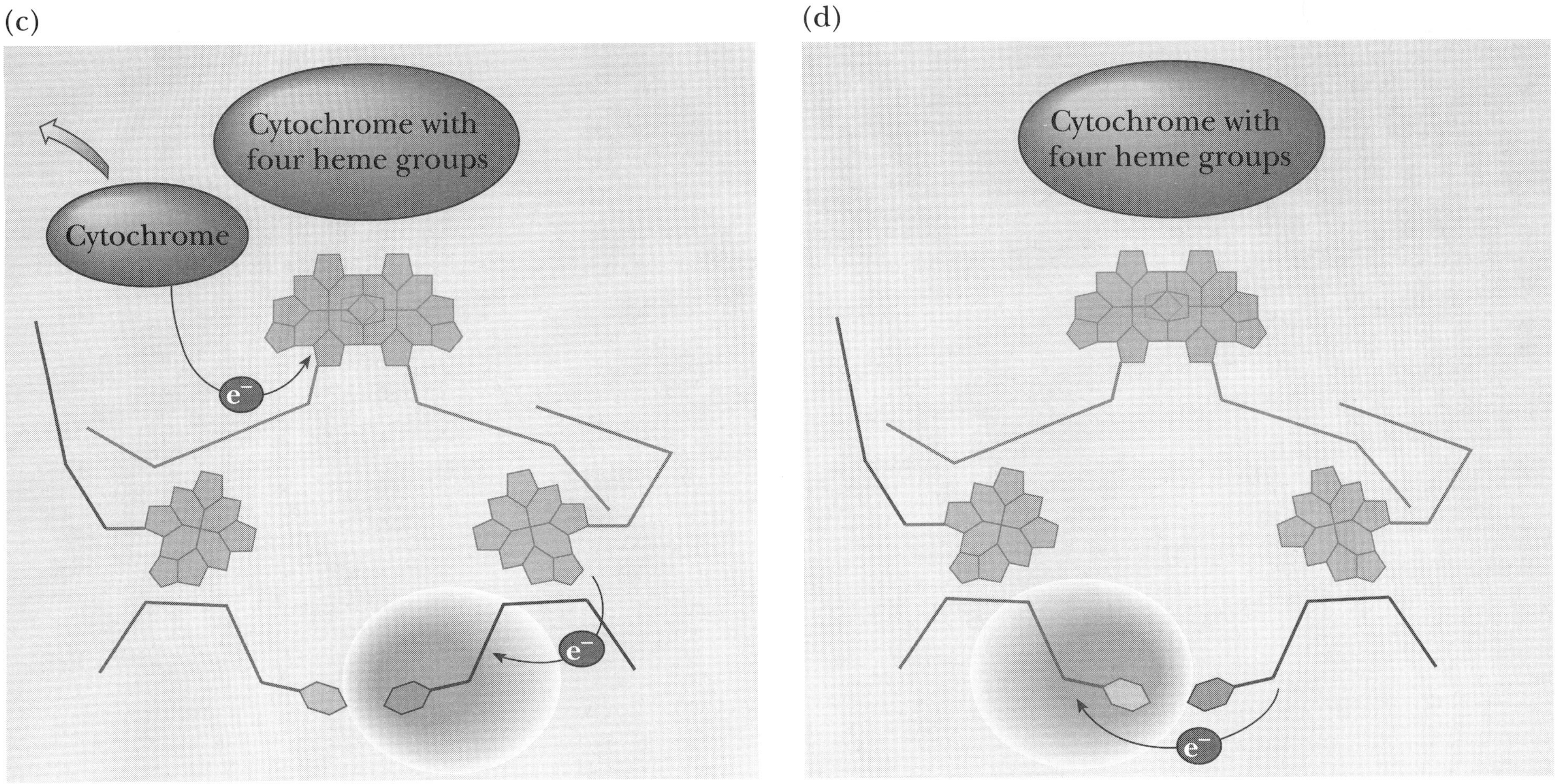

***Figure 19.9cd Molecular events that take place at the photosynthetic reaction center of* Rhodopseudomonas**

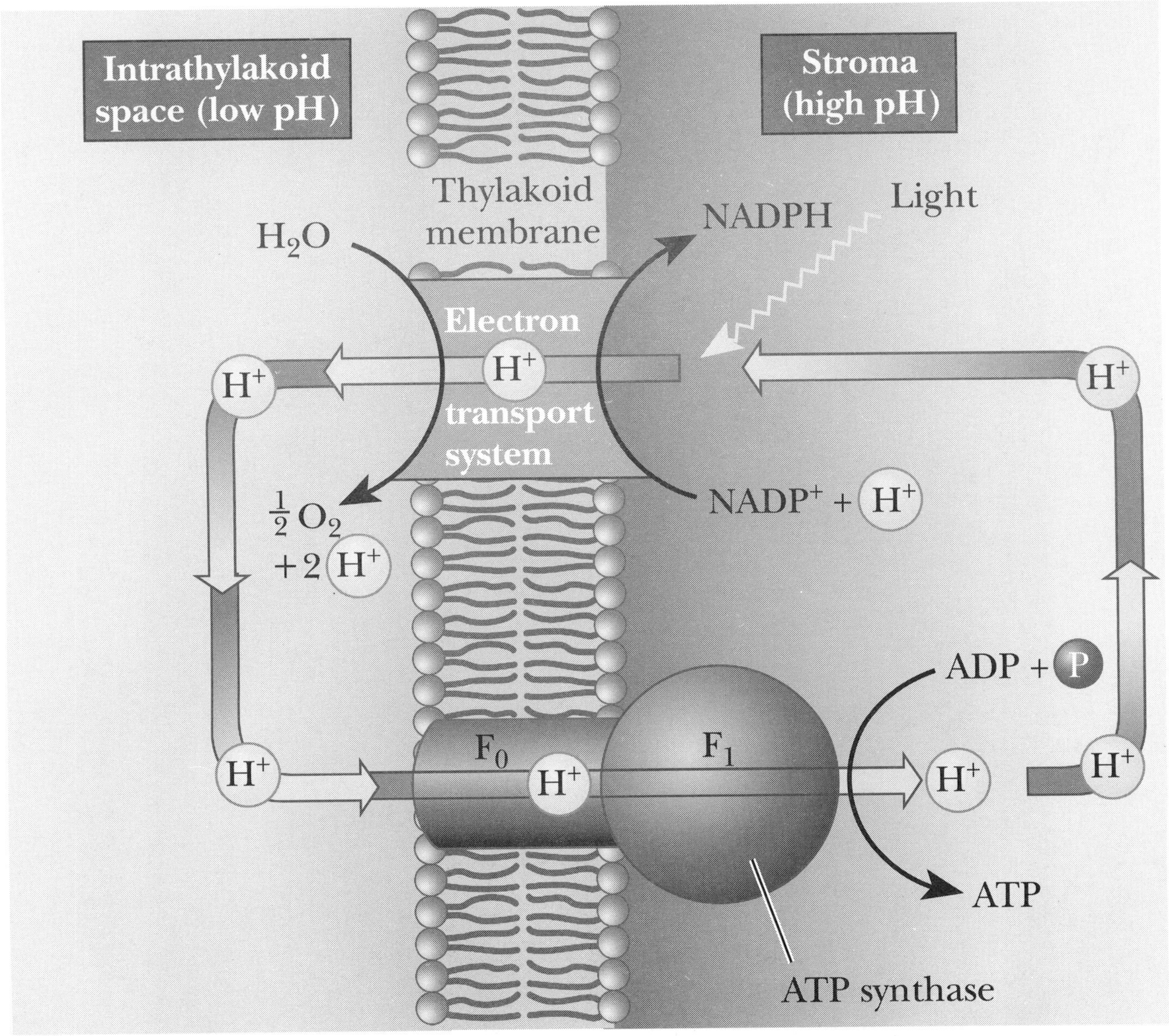

Figure 19.11 The relationship between photophosphorylation and the proton gradient

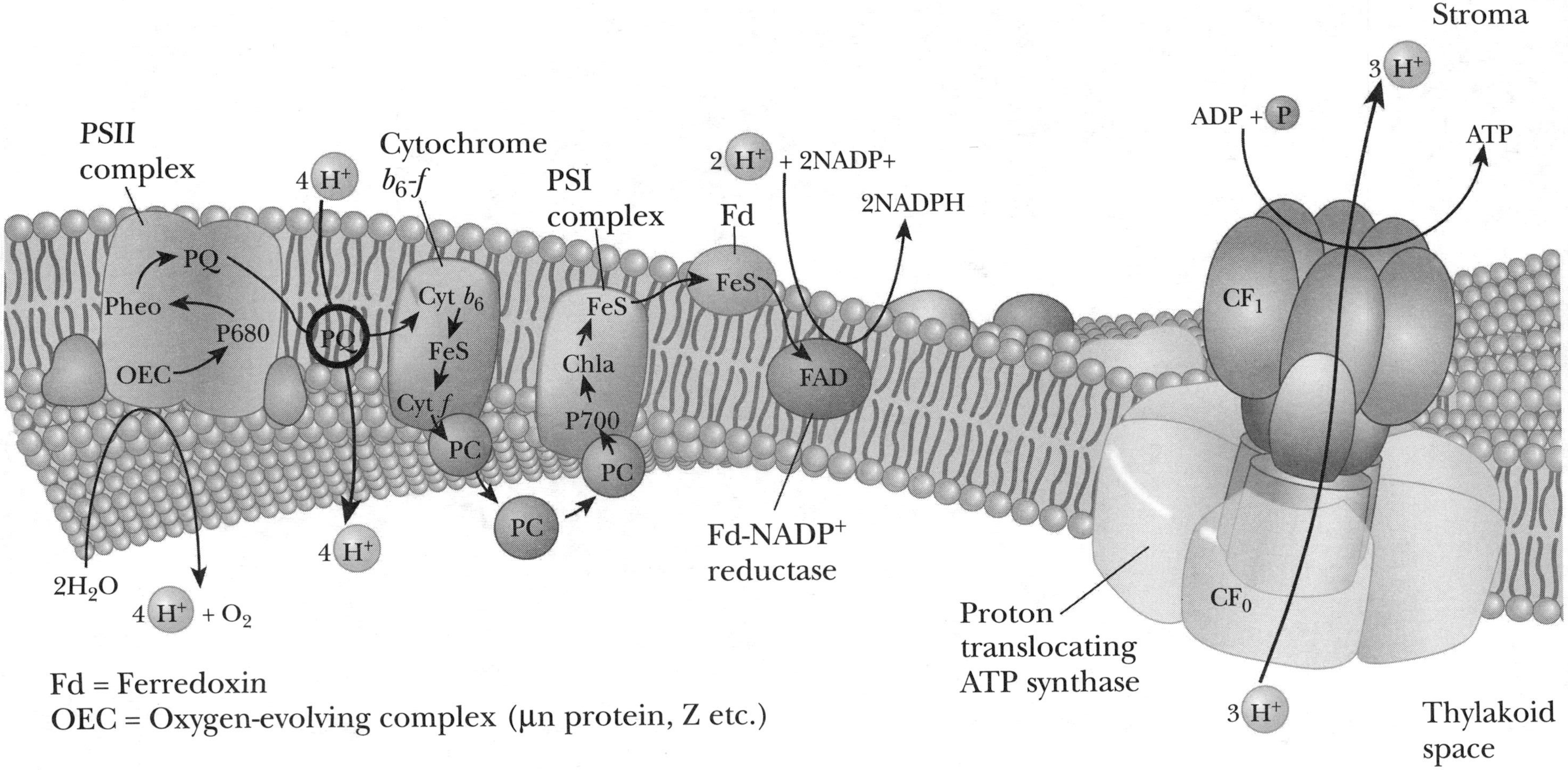

Figure 19.12 The components of the electron transport chain of the thylakoid membrane

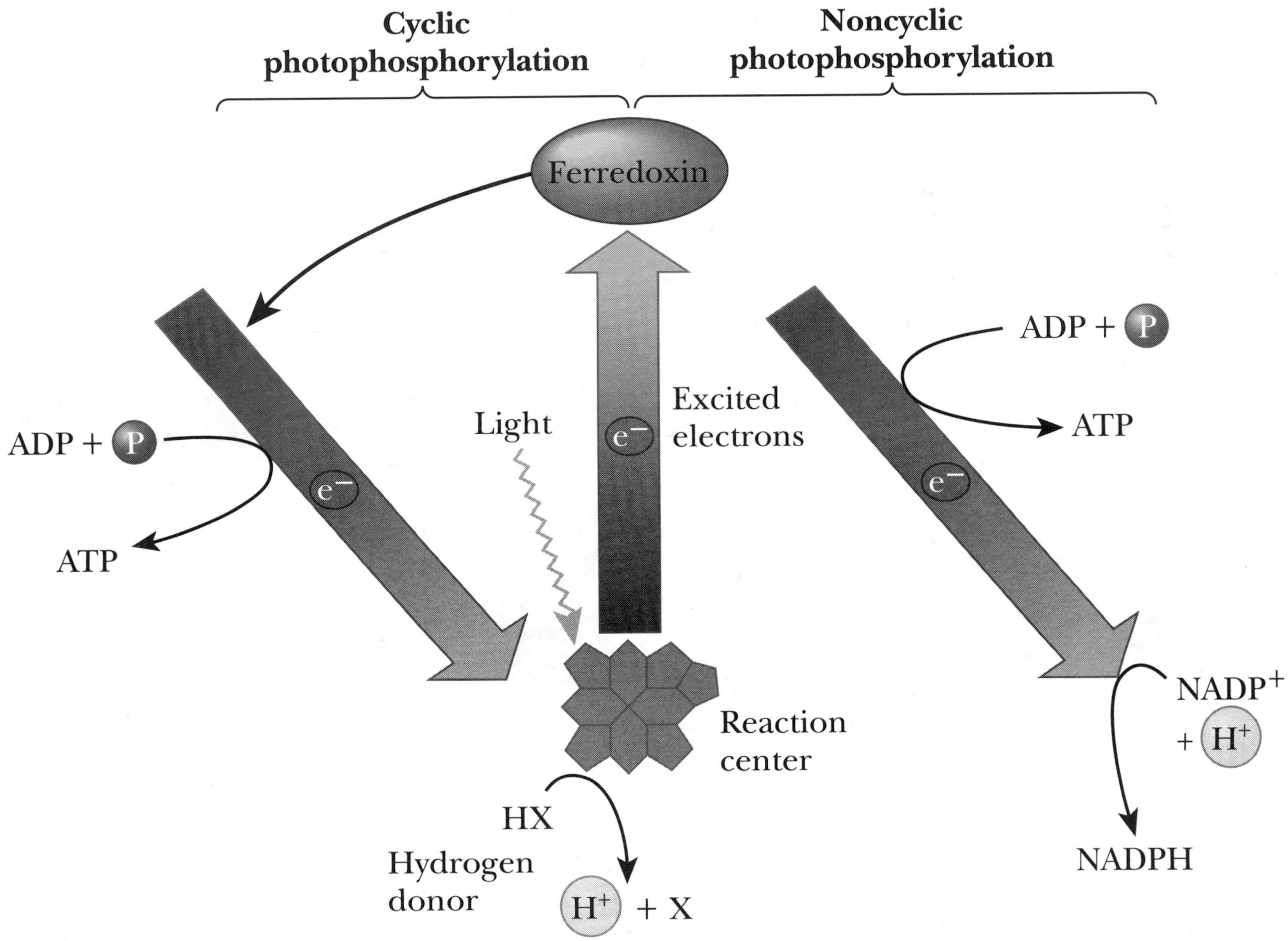

Figure 19.13 The two possible electron transfer pathways in a photosynthetic anaerobe

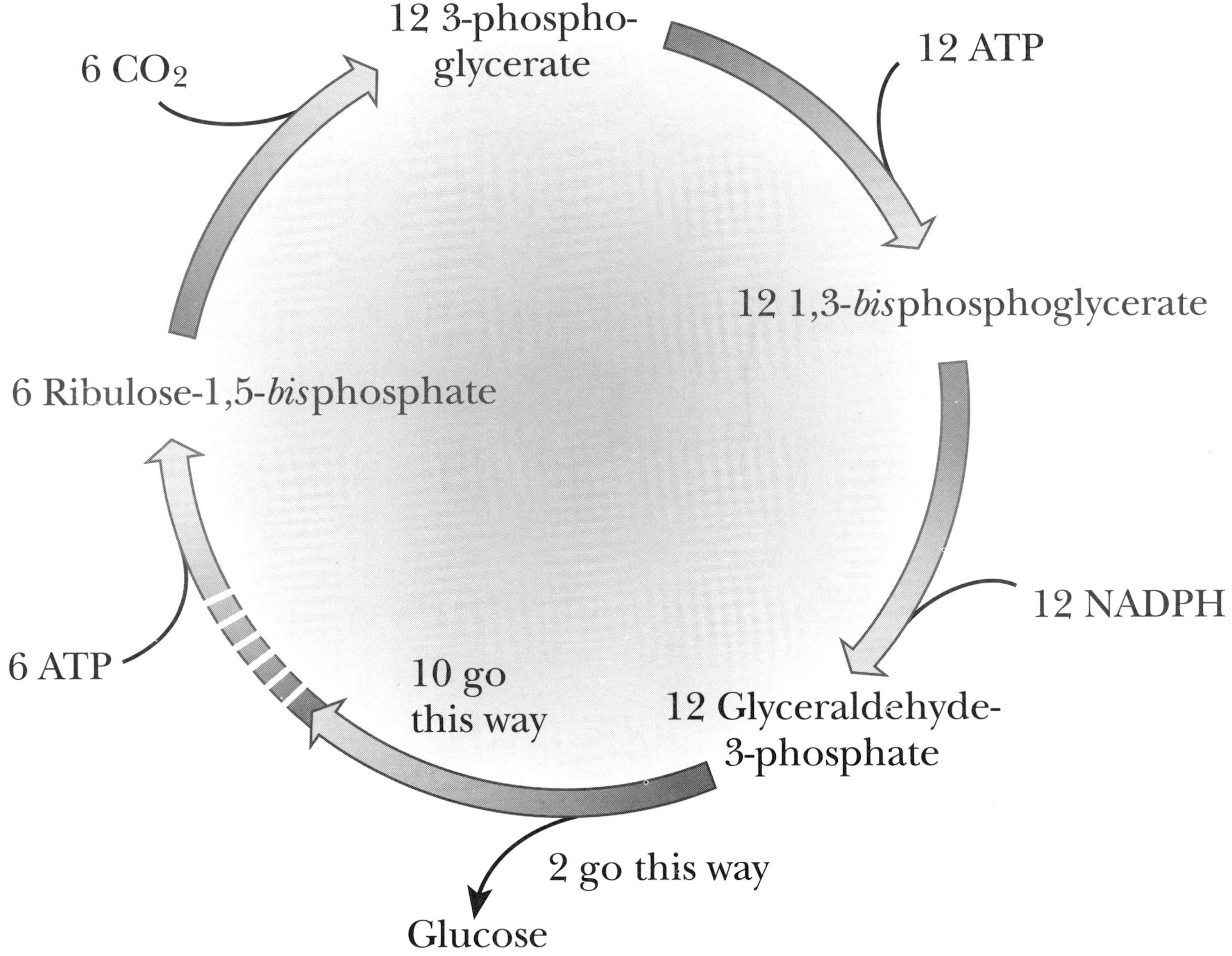

Figure 19.14 The main features of the Calvin cycle

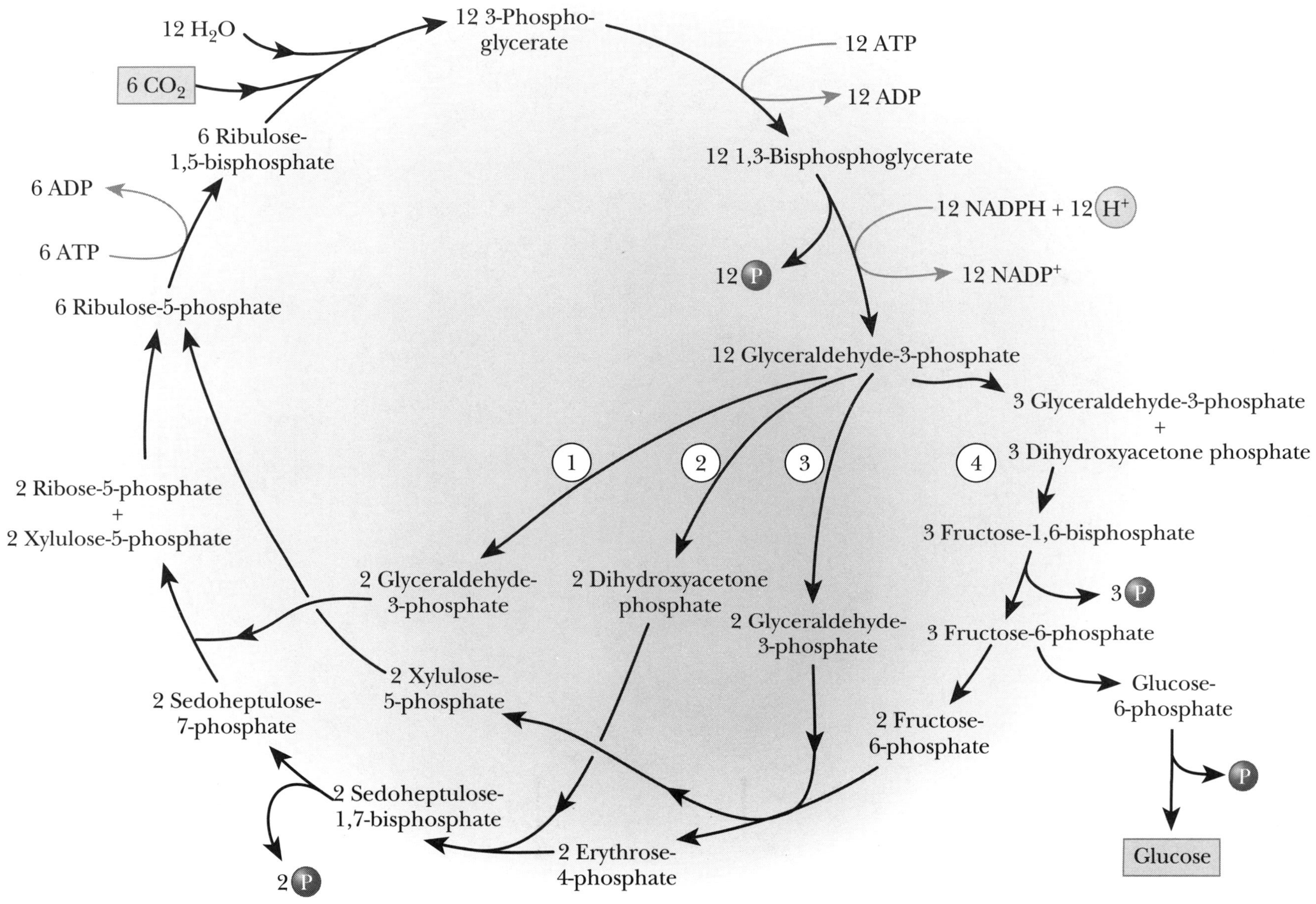

12 H_2O
6 CO_2
12 3-Phospho-glycerate
12 ATP
12 ADP
12 1,3-Bisphosphoglycerate
12 NADPH + 12 H^+
12 P
12 $NADP^+$
12 Glyceraldehyde-3-phosphate
3 Glyceraldehyde-3-phosphate + 3 Dihydroxyacetone phosphate
1
2
3
4
3 Fructose-1,6-bisphosphate
3 P
3 Fructose-6-phosphate
Glucose-6-phosphate
P
Glucose
2 Fructose-6-phosphate
2 Glyceraldehyde-3-phosphate
2 Dihydroxyacetone phosphate
2 Glyceraldehyde-3-phosphate
2 Xylulose-5-phosphate
2 Erythrose-4-phosphate
2 Sedoheptulose-1,7-bisphosphate
2 P
2 Sedoheptulose-7-phosphate
2 Ribose-5-phosphate + 2 Xylulose-5-phosphate
6 Ribulose-5-phosphate
6 ATP
6 ADP
6 Ribulose-1,5-bisphosphate

Figure 19.20 The complete Calvin cycle

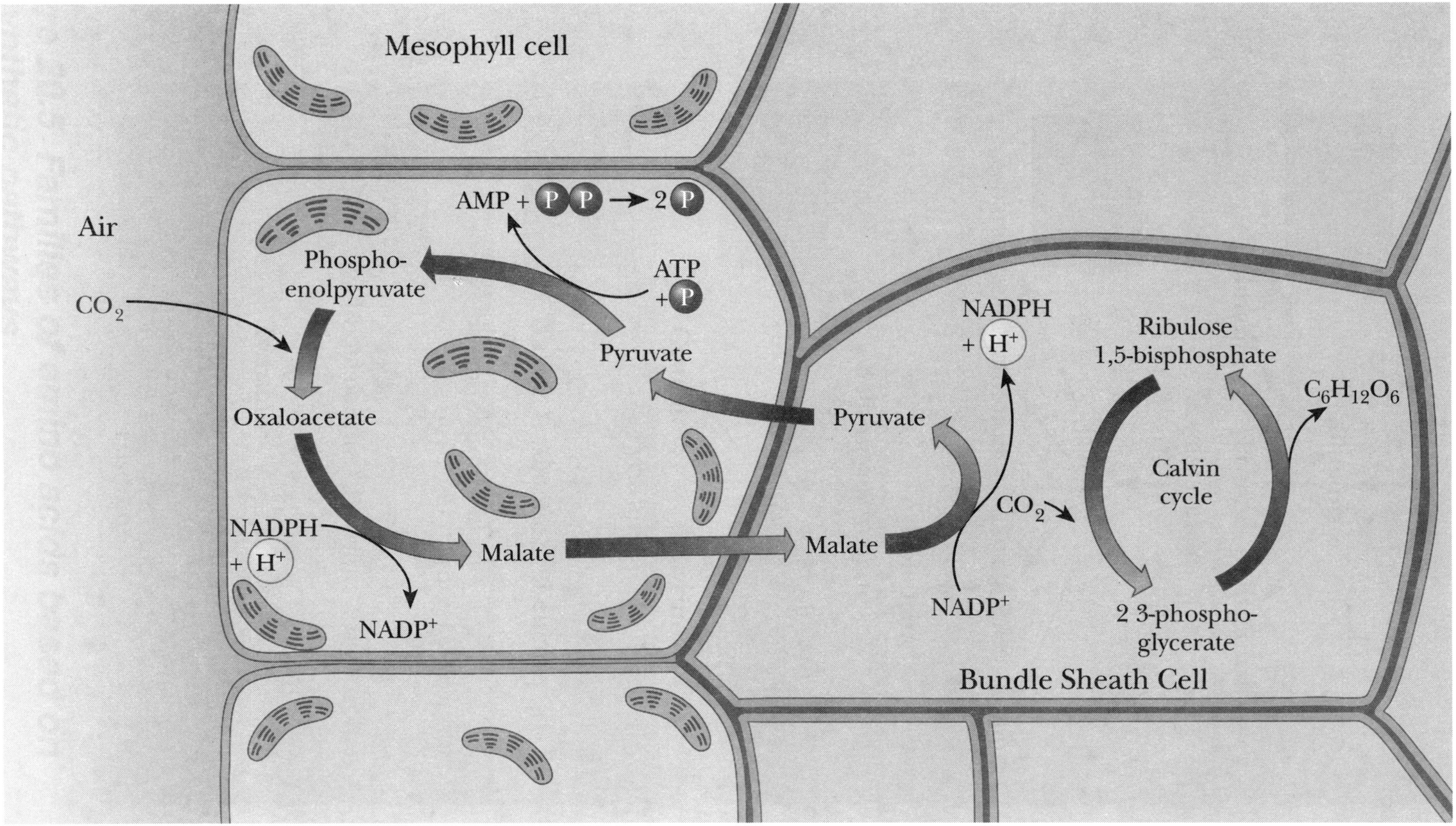
Mesophyll cell
Air
CO_2
AMP + P P → 2 P
ATP
+ P
Phospho-
enolpyruvate
Pyruvate
Oxaloacetate
NADPH
+ H^+
$NADP^+$
Malate
Malate
Pyruvate
NADPH
+ H^+
CO_2
$NADP^+$
Ribulose
1,5-bisphosphate
$C_6H_{12}O_6$
Calvin
cycle
2 3-phospho-
glycerate
Bundle Sheath Cell

Figure 19.21 The C_4 pathway

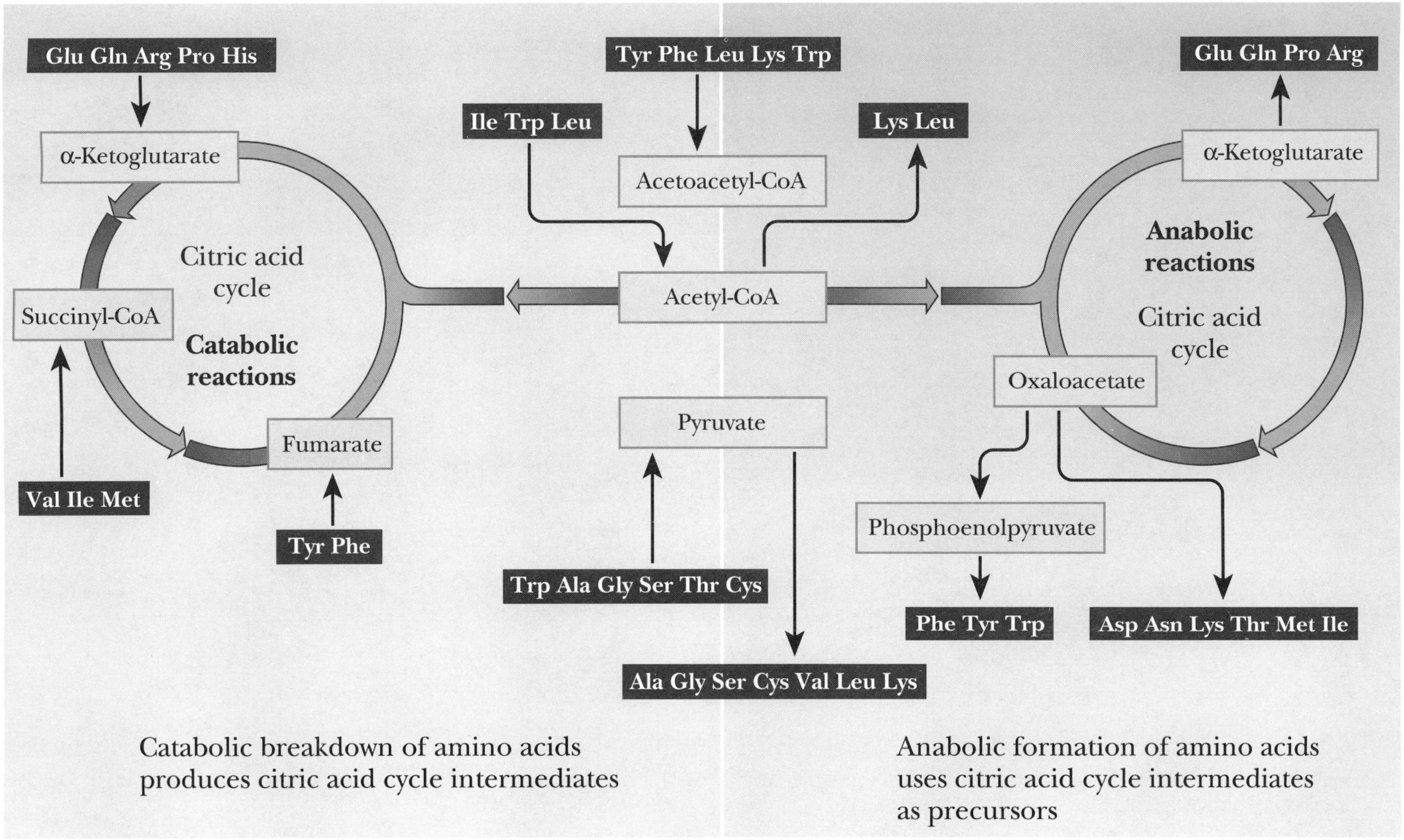

Figure 20.6 The relationship between amino acid metabolism and the citric acid cycle

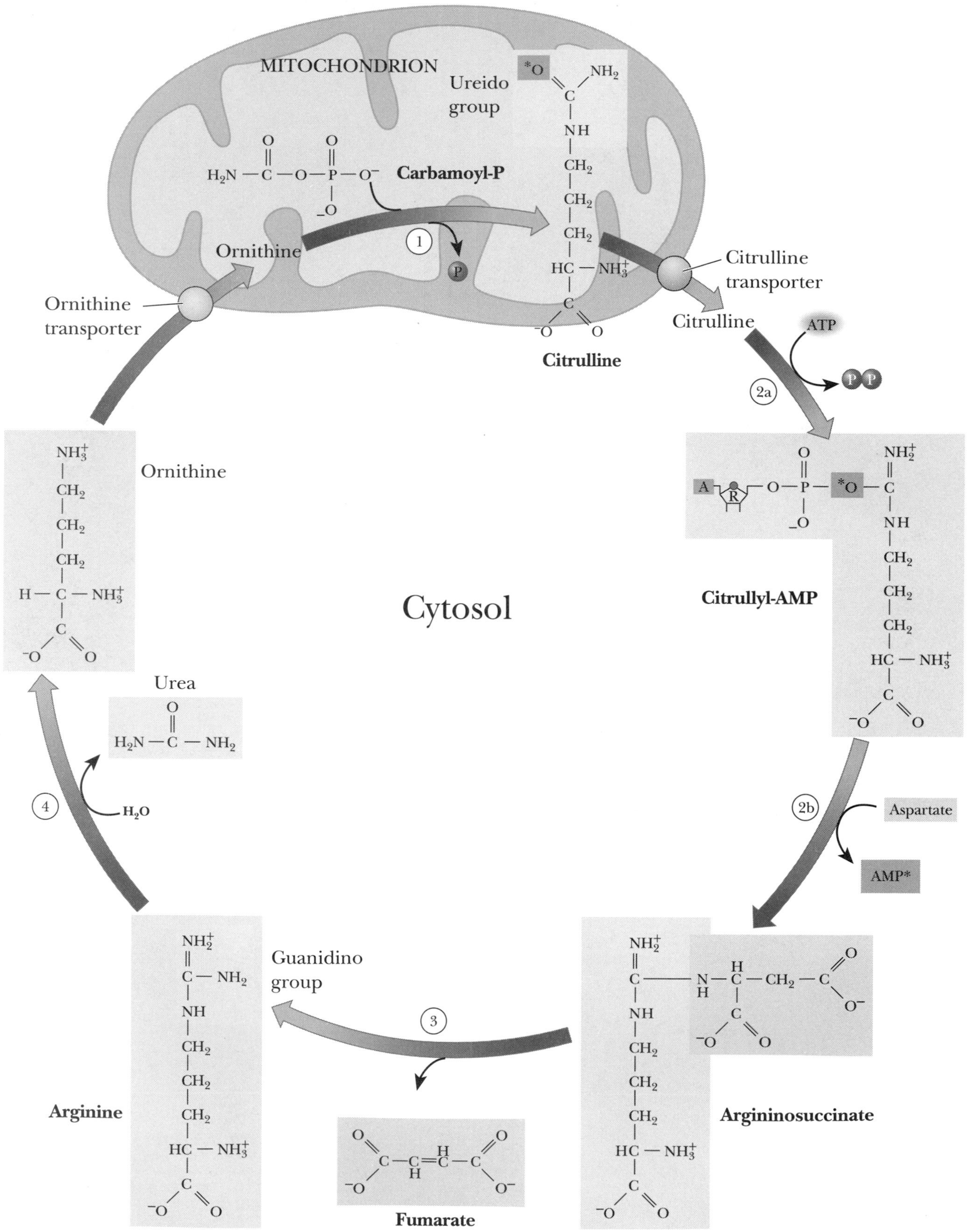
MITOCHONDRION
Ureido group
Carbamoyl-P
Ornithine
Citrulline
Citrulline transporter
Citrulline
Ornithine transporter
ATP
2a
Citrullyl-AMP
Ornithine
Cytosol
Urea
4
H_2O
2b
Aspartate
AMP*
Guanidino group
3
Arginine
Fumarate
Argininosuccinate

Figure 20.17 The urea cycle

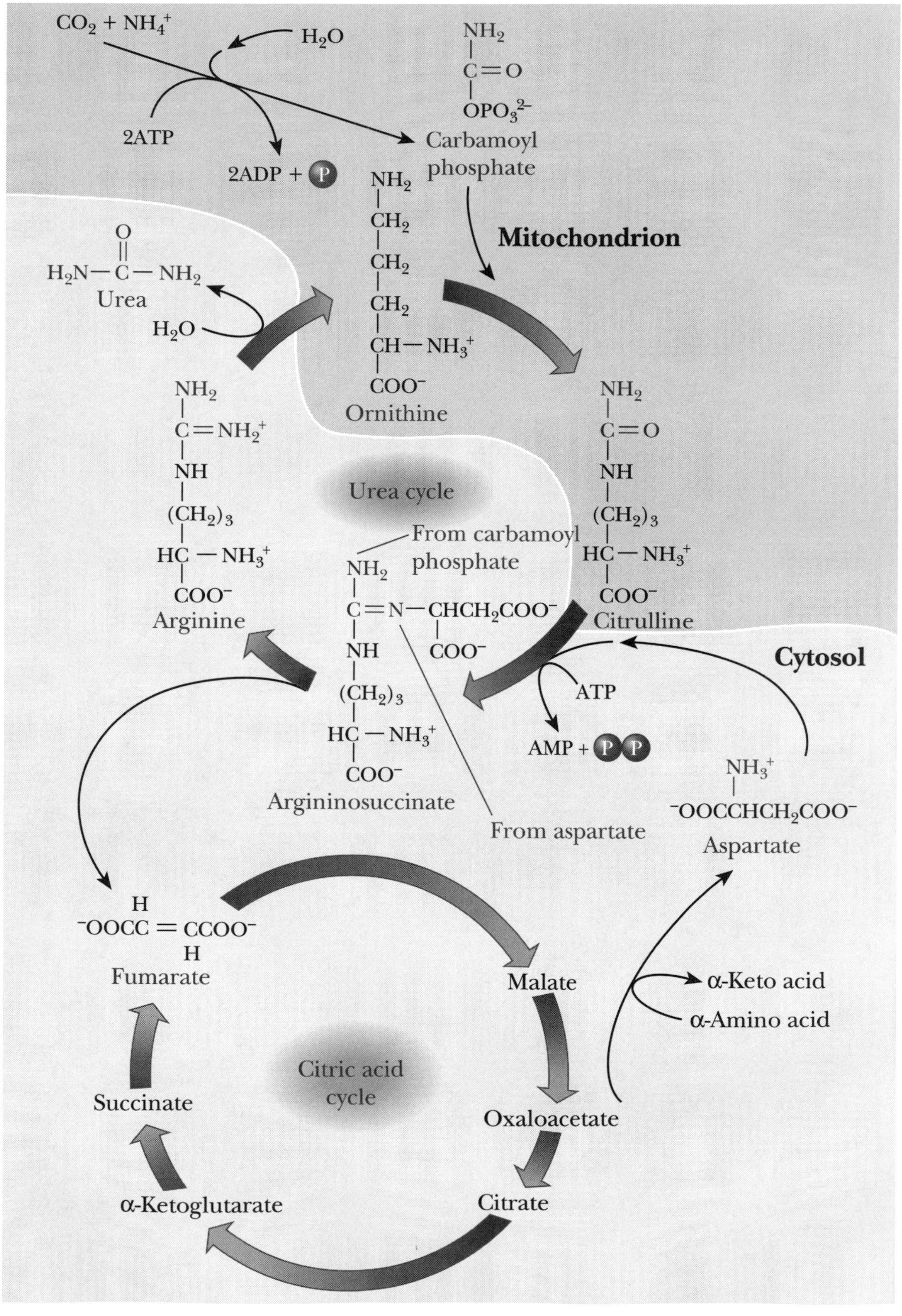

Figure 20.18 The urea cycle and some of its links to the citric acid cycle

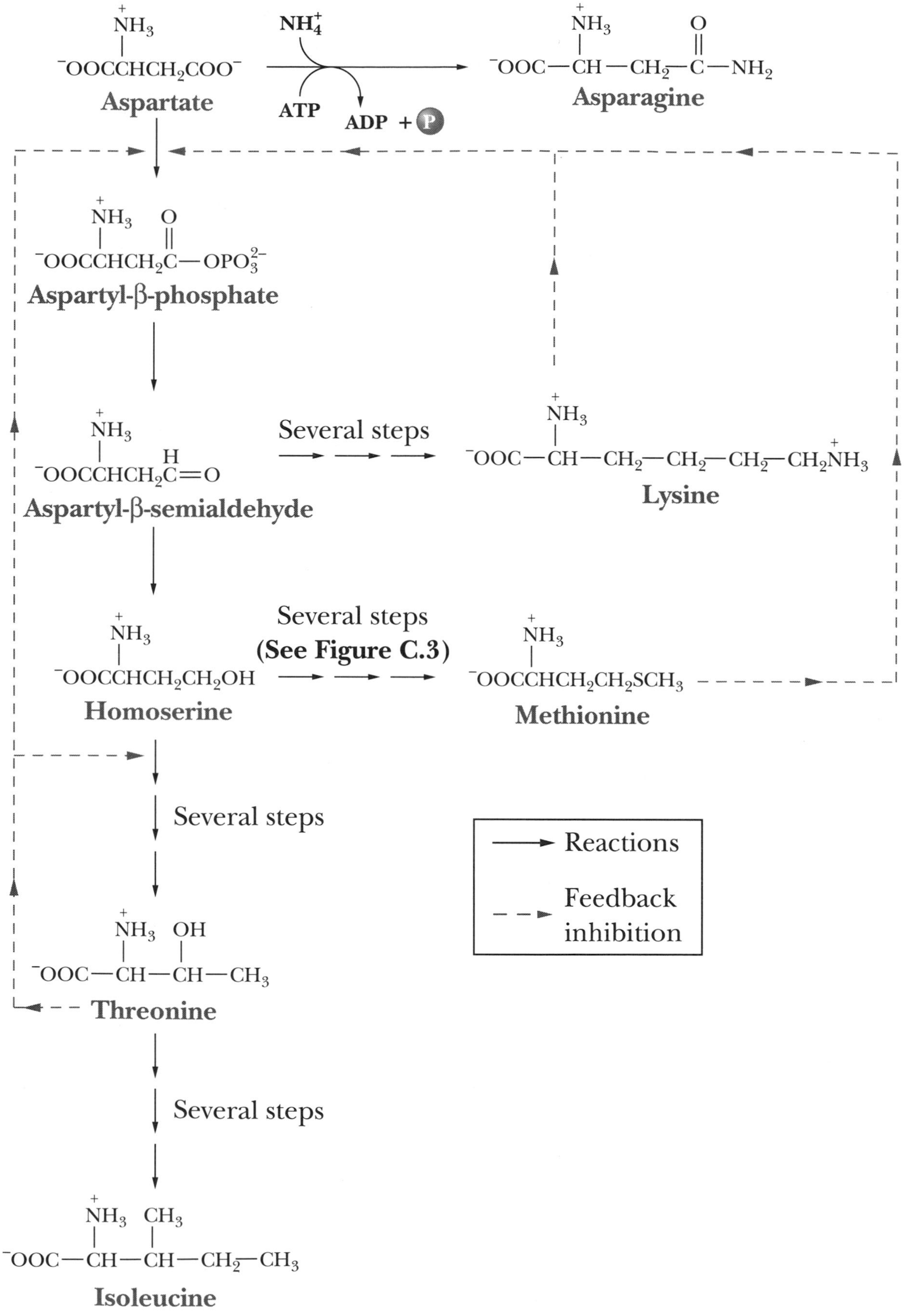

Figure C.4 Feedback control in the biosynthesis of amino acids of the aspartate family

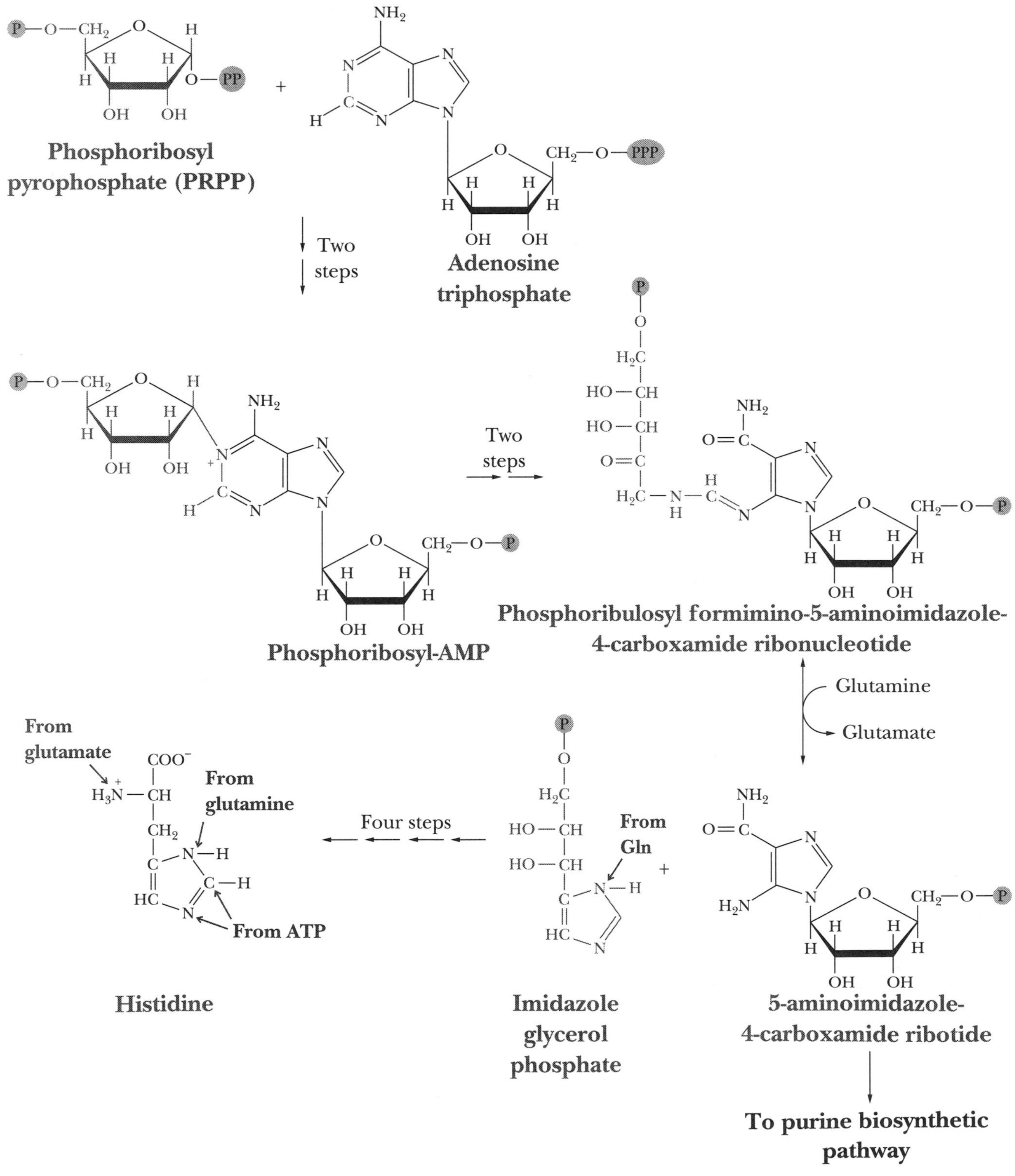

Phosphoribosyl pyrophosphate (PRPP)
Adenosine triphosphate
Two steps
Phosphoribosyl-AMP
Two steps
Phosphoribulosyl formimino-5-aminoimidazole-4-carboxamide ribonucleotide
Glutamine
Glutamate
5-aminoimidazole-4-carboxamide ribotide
To purine biosynthetic pathway
Imidazole glycerol phosphate
From Gln
Four steps
Histidine
From glutamate
From glutamine
From ATP

Figure C.9 Histidine biosynthesis

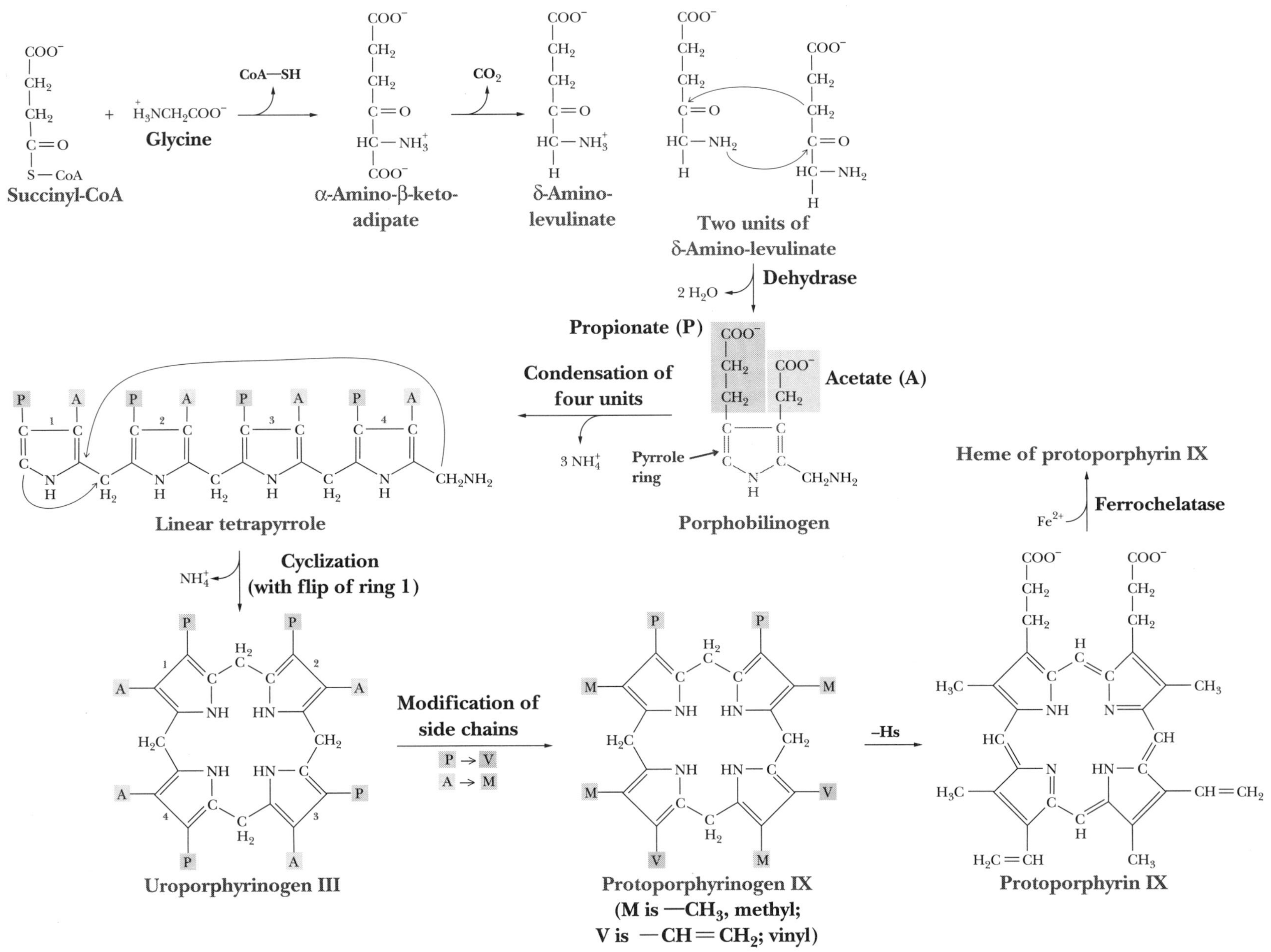

Succinyl-CoA
Glycine
CoA—SH
α-Amino-β-keto-adipate
CO_2
δ-Amino-levulinate
Two units of δ-Amino-levulinate
Dehydrase
$2 H_2O$
Propionate (P)
Acetate (A)
Pyrrole ring
Porphobilinogen
Condensation of four units
$3 NH_4^+$
Linear tetrapyrrole
Cyclization (with flip of ring 1)
NH_4^+
Uroporphyrinogen III
Modification of side chains
P → V
A → M
Protoporphyrinogen IX (M is $—CH_3$, methyl; V is $—CH=CH_2$; vinyl)
–Hs
Protoporphyrin IX
Ferrochelatase
Fe^{2+}
Heme of protoporphyrin IX

Figure C.10 The biosynthesis of porphyrins

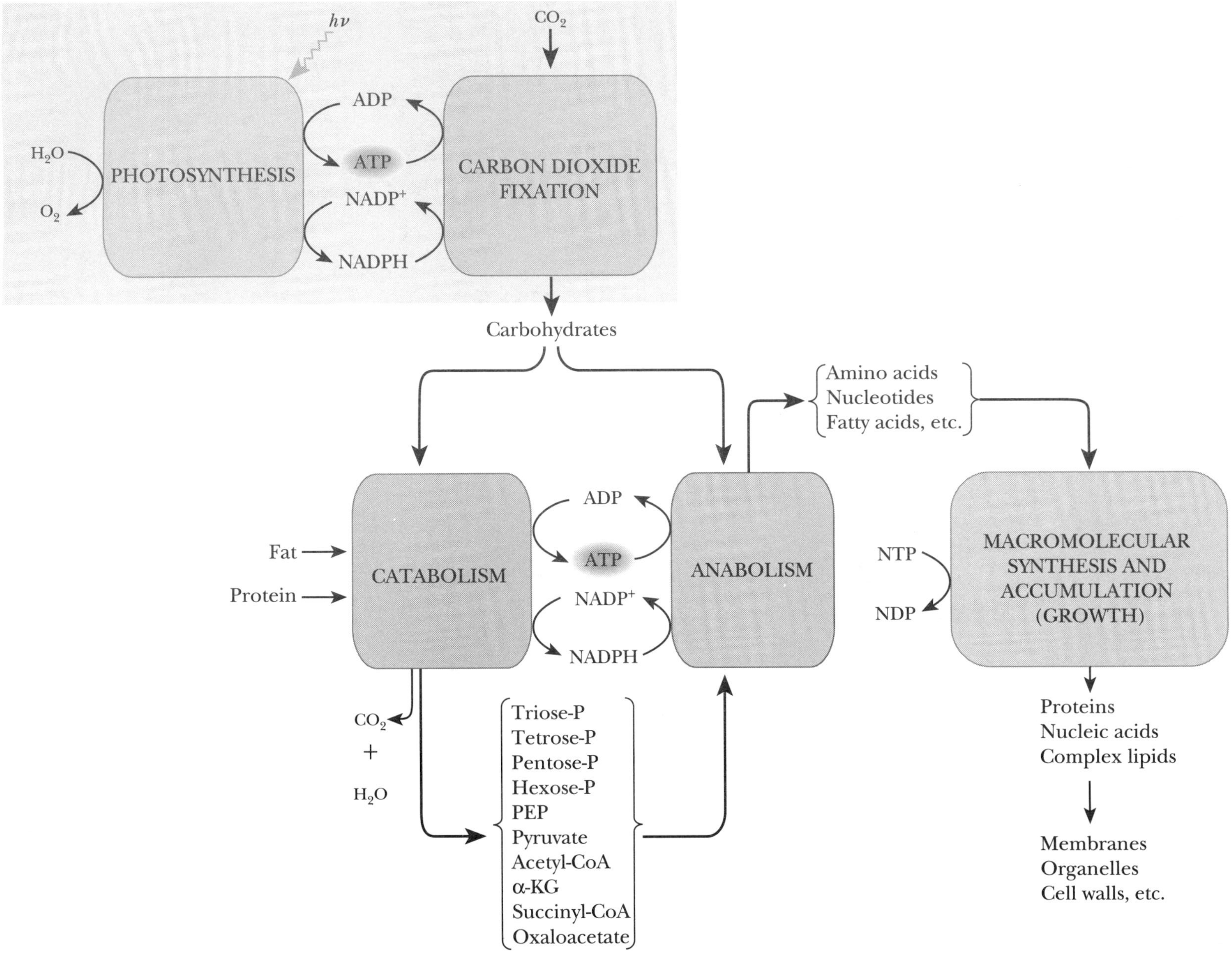

Figure 21.1 Block diagram of intermediary metabolism

Central nervous system
Feedback
Feedback
Hypothalamus
Nerve impulses
CRF
TRF
GnRF
GRF
Anterior pituitary
Posterior pituitary
ACTH
TSH
FSH
LH
Growth Hormone
Vasopressin
Oxytocin
Adrenal cortex
Thyroid
Testes/Ovary
Adreno-cortical hormones
Thyroxine
Androgens
Estrogens
Muscles, liver
Sex accessory tissues
Bone
Water balance
Lactation, uterine contraction

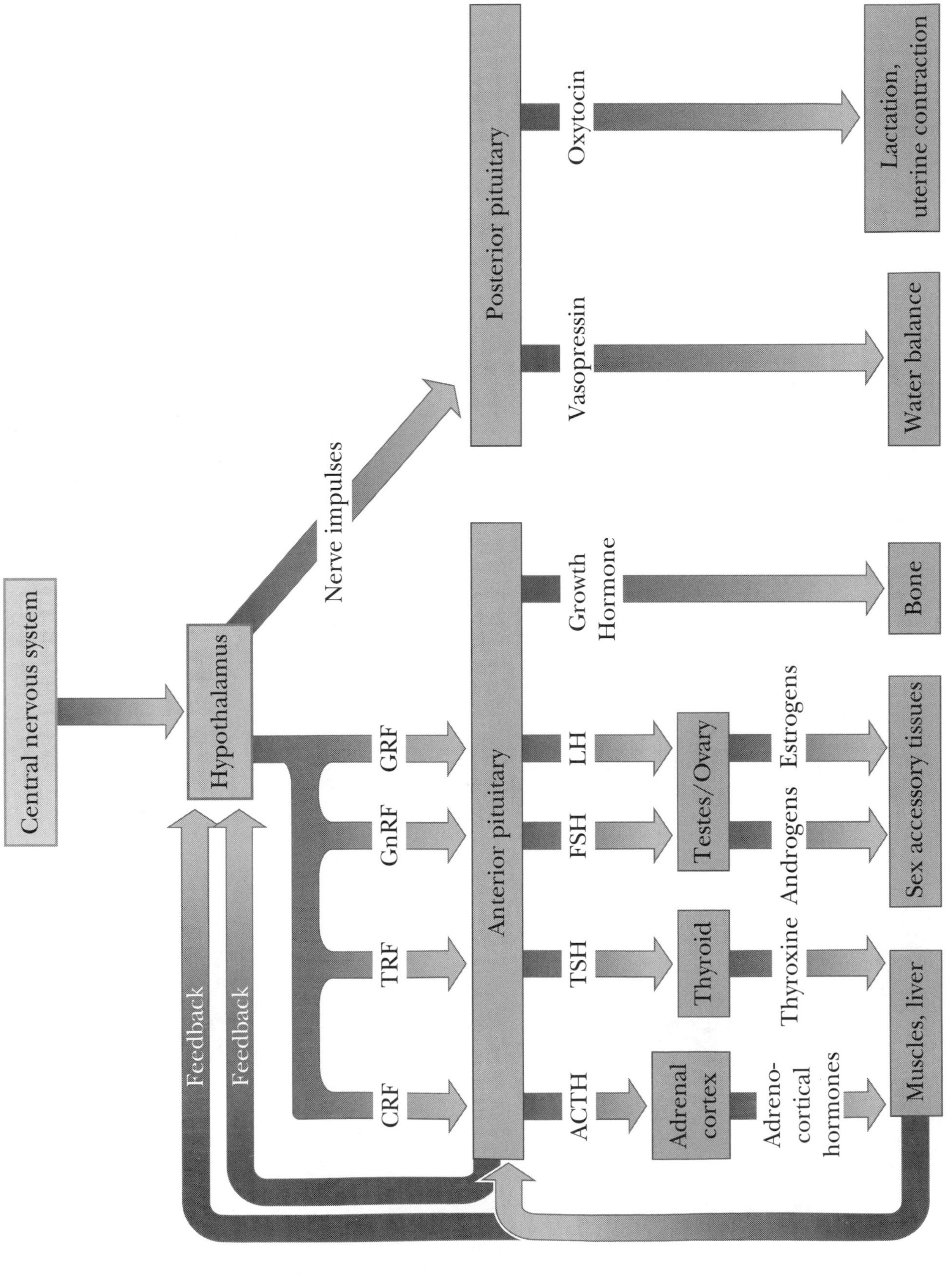

Figure 21.4 Hormonal control system

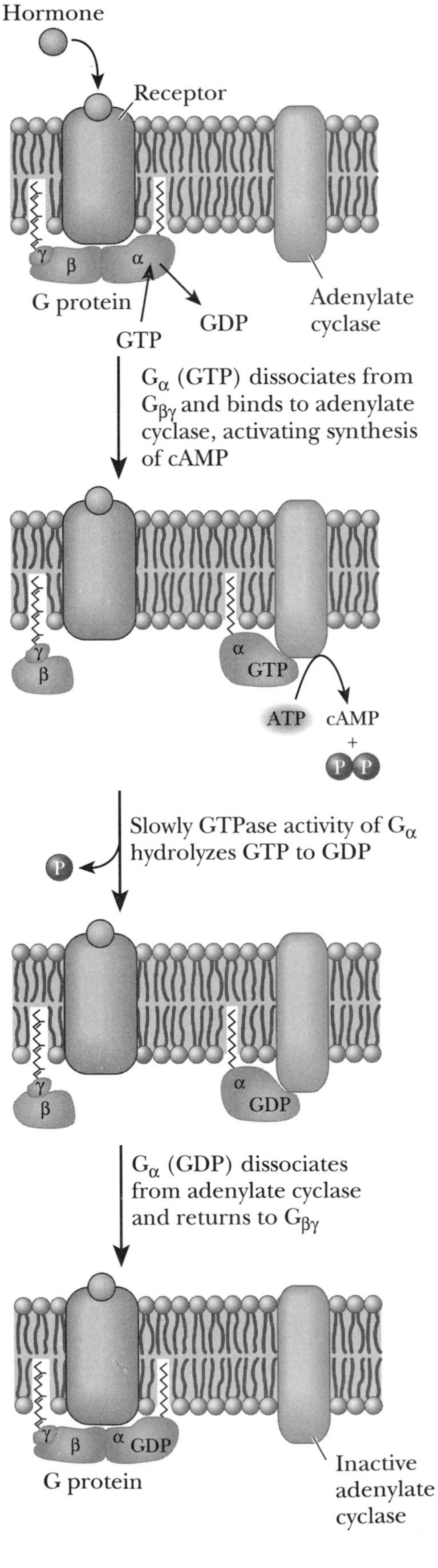

Figure 21.5 Activation of adenylyate cyclase by heterodimeric G proteins

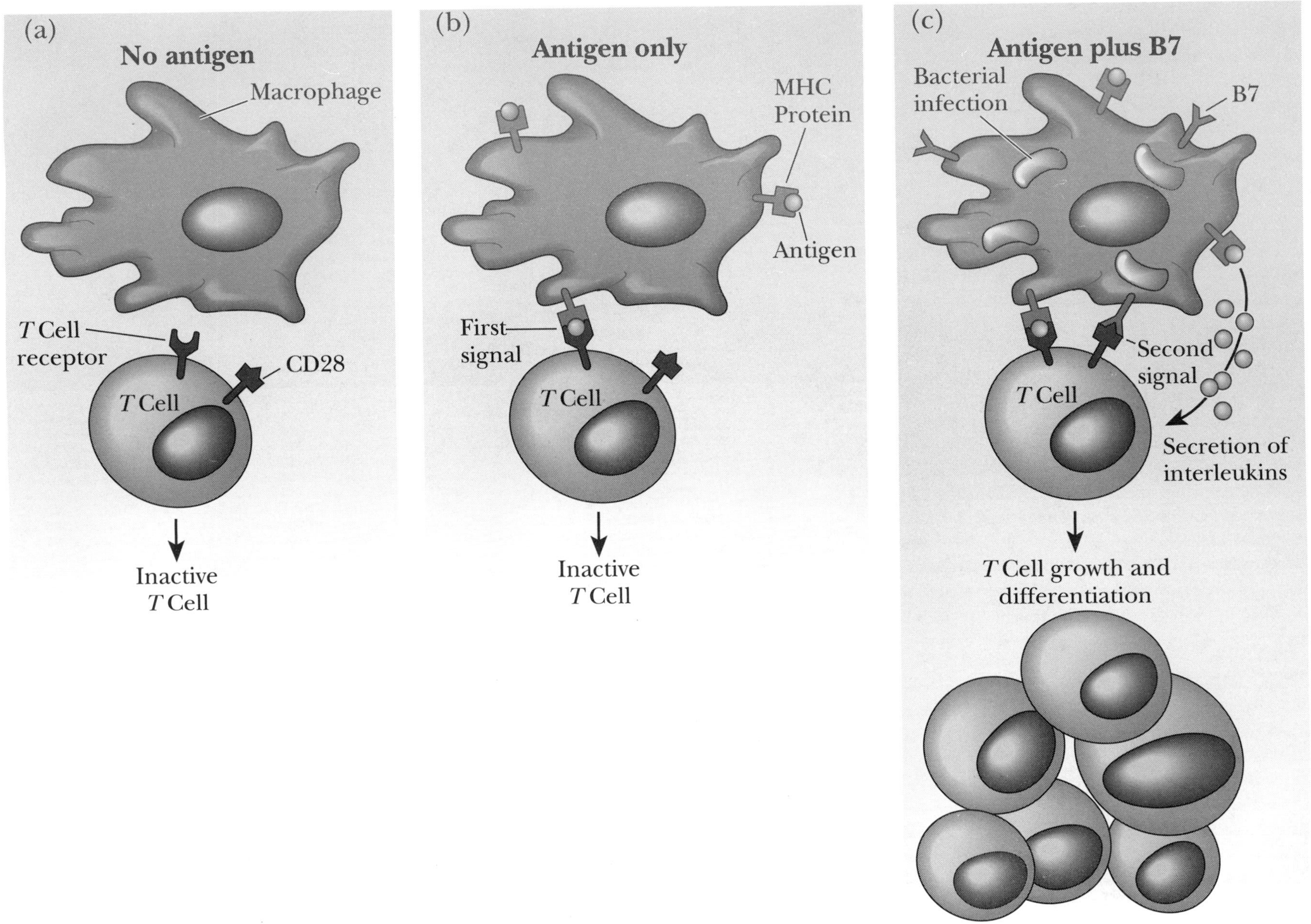

Figure 21.12 A two-stage process leads to the growth and differentiation of T cells